8th Roma International Conference on AstroParticle Physics (RICAP-22)

EPJ Web of Conferences Volume 280 (2023)

Roma, Italy
6 – 9 September 2022

Editors:

Antonio Capone
Mario De Vincenzi
Aldo Morselli

ISBN: 978-1-7138-7249-8

Printed from e-media with permission by:

Curran Associates, Inc.
57 Morehouse Lane
Red Hook, NY 12571

Some format issues inherent in the e-media version may also appear in this print version.

For additional information, please contact EDP Sciences – Web of Conferences
at the address below.

EDP Sciences – Web of Conferences
17, Avenue du Hoggar
Parc d'Activité de Courtabœuf
BP 112
F-91944 Les Ulis Cedex A
France

Phone: +33 (0) 1 69 18 75 75
Fax: +33 (0) 1 69 28 84 91

contact-edps@webofconferences.org

Additional copies of this publication are available from:

Curran Associates, Inc.
57 Morehouse Lane
Red Hook, NY 12571 USA
Phone: 845-758-0400
Fax: 845-758-2633
Email: curran@proceedings.com
Web: www.proceedings.com

TABLE OF CONTENTS

edp sciences EPJ Web of Conferences

Statement of Peer review

In submitting conference proceedings to EPJ Web of Conferences, I certify to the Publisher that I adhere to the **Policy on Publishing Integrity** of the journal in order to safeguard good scientific practice in publishing.

1. All articles have been subjected to peer review administered by the proceedings editors.
2. Reviews have been conducted by expert referees, who have been requested to provide unbiased and constructive comments aimed, whenever possible, at improving the work.
3. Proceedings editors have taken all reasonable steps to ensure the quality of the materials they publish and their decision to accept or reject a paper for publication has been based only on the merits of the work and the relevance to the journal.

Roma International Astroparticle Physics Conference, VIII edition, Roma, Italy, September 6 – 9, 2022

Proceedings editor(s): Antonio Capone, Mario De Vincenzi, Aldo Morselli

Date and editor's signature March 6th, 2023

EPJ Web of Conferences **280**, 01001 (2023) https://doi.org/10.1051/epjconf/202328001001
RICAP-22

Latest results from the DAMPE space mission

Irene Cagnoli[1,2], *Ivan* De Mitri[1,2*], and *Leandro* Silveri[1,2]
on behalf of the DAMPE collaboration

[1]Gran Sasso Science Institute (GSSI), L'Aquila, Italy
[2]Istituto Nazionale di Fisica Nucleare (INFN)—Laboratori Nazionali del Gran Sasso, L'Aquila, Italy

Abstract. The space-based DAMPE (DArk Matter Particle Explorer) particle detector has been taking data for more than 7 years since its successful launch in December 2015. Its main scientific goals include the indirect search for dark matter signatures in the cosmic lepton and gamma-ray spectra, the study of galactic cosmic rays up to energies of hundreds of TeV and studies on high-energy gamma ray astronomy. The measurement of galactic cosmic ray spectra are reported here, those being fundamental tools to investigate the mechanisms of acceleration at their sources and propagation through the interstellar medium. Results on proton and helium, which revealed new spectral features, are described. Ongoing analyses on the cosmic ray light, medium and heavy mass nuclei are outlined, together with studies of the so-called secondary cosmic rays. Latest results on gamma-ray astronomy and dark matter search will be also summarized.

1 Introduction

The DArk Matter Particle Explorer [1] (DAMPE) is a space particle detector, launched on 17 December 2015 by a collaboration of Chinese, Italian and Swiss scientific institutions. It is operating in a sun-synchronous orbit at an altitude of ~500 km, with an inclination of ~97° and a period of about 95 minutes. The total mass of the payload is about 1400 kg with a power consumption of about 400 W.

The main scientific objectives of the DAMPE space mission are the indirect search for dark matter (DM) signatures, the high energy gamma ray astronomy, the study of cosmic electron + positron spectrum and the study of cosmic ray protons and nuclei. Therefore, the DAMPE detector is designed to have excellent performances for the detection of cosmic electrons and photons in the energy range from 10 GeV to 10 TeV and to measure the fluxes and the mass composition of galactic cosmic ray nuclei from 50 GeV up to hundreds of TeV. In particular, it is characterised by a large acceptance, corresponding to 0.3 m^2sr for electrons above 30 GeV, and an accurate angular resolution for electrons and photons that is less than 0.2° with ~0.1 sr of field of view. Moreover, it has an excellent energy resolution, being 1.2% for electrons at 100 GeV and below 40% for nuclei at 800 GeV [1].

*e-mail: ivan.demitri@gssi.it

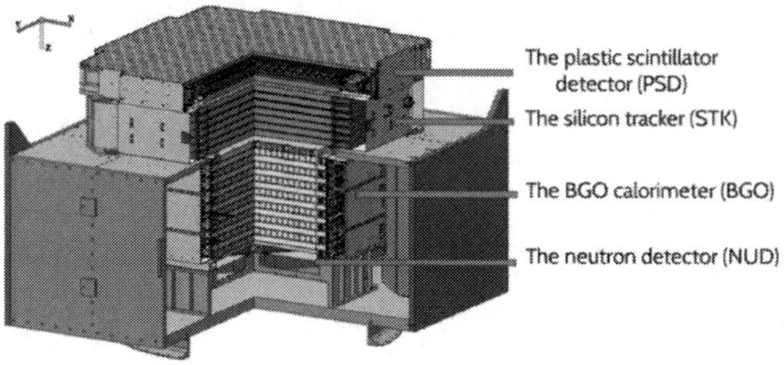

Figure 1: Schematic view of the DAMPE detector with subdetector modules.

2 The DAMPE detector

The detector consists of four subdetector modules which are, from top to bottom, a Plastic Scintillator Detector (PSD), a Silicon-Tungsten tracKer converter (STK), a BGO calorimeter and a NeUtron Detector (NUD), shown in figure 1.

The PSD [2] is designed to measure the absolute value of the charge of incident cosmic rays (CRs) up to iron and acts also as an anti-coincidence detector for γ-ray observations. It is composed of 2 planes, oriented respectively in x and y directions, having 41 plastic scintillator bars each. The bars are disposed in 2 staggered layers to increase the hermeticity and readout by 2 photomultiplier tubes (PMTs).

The STK [3] operates to reconstruct the trajectory of the incident particles. It has 6 x-y tracking planes of silicon microstrip detectors that provide independent particle coordinate measurements in the two orthogonal directions perpendicular to the telescope pointing direction. To promote the photon conversion, 3 tungsten layers (1 mm thick) are inserted after the first, second and third tracking planes.

The calorimeter [4] is made of 14 layers of bismuth germanium oxide (BGO) bars and it is characterised by a total depth of ~32 radiation lengths and ~1.6 nuclear interaction lengths. It is designed to measure the energy of incident particles and to provide efficient electromagnetic/hadronic showers discrimination. For this reason the layers are in a hodoscopic configuration to perform a 3D imaging of the shower profiles and they are composed of 22 BGO bars (2.5 x 2.5 x 60 cm^3) readout by PMTs on the two ends.

The NUD [5] detects the secondary neutrons within few μs after their production in the BGO, thus improving the electron/hadron separation power.

2.1 Beam tests at CERN and on-orbit calibration

Several beam test campaigns had been performed at CERN, between 2014 and 2015 [1] to validate the detector performances and to determine the calibration parameters. Different particle beams had been used, namely electrons, photons, pions, protons and nuclear fragments. Then, in the first two weeks after the DAMPE launch a dedicated calibration was performed using cosmic rays and the results were compared with those obtained in the beam tests at CERN to verify the on-orbit performance of the detector. Furthermore, in order to ensure good quality of the energy and charge measurements for individual events, the detection efficiency and to check the stability of the detector, a calibration of each sub-detector is periodically performed during the on-orbit operation [6]. In particular, twice per orbit the DAQ system collects 40 s of data for the electronic linearity and the pedestal calibration.

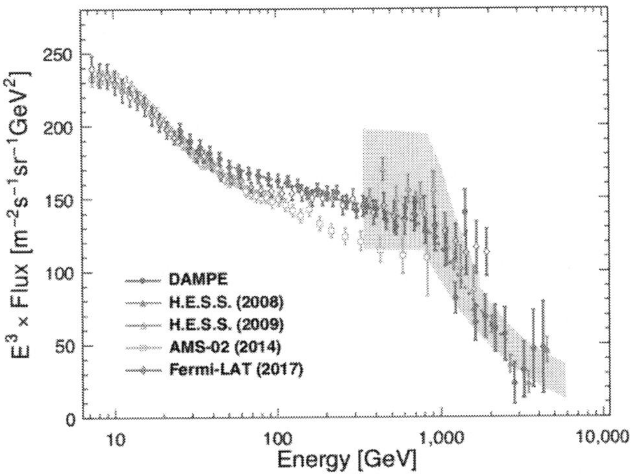

Figure 2: DAMPE CRE spectrum (red markers) multiplied by E^3, compared with measurements from other experiments. The error bars include both the systematic and the statistical uncertainties added in quadrature. The red dashed line represents the smoothly broken power law fit [7].

While, other calibrations, including the minimum ionization particle response and detector alignment, are performed using the nominal science data.

3 Results and ongoing analyses

3.1 All-electron (e^++e^-) spectrum

One of the main goals of the experiment is the measurement of the cosmic ray electron plus positron (CRE) spectrum, since it dispenses a probe for the study of Galactic high energy processes and it could enable the indirect dark matter observation. DAMPE performed this measurement between 25 GeV and 4.6 TeV using 530 days of data [7]. The electron-positron sample was selected with a high purity thanks to the efficient electromagnetic/hadronic shower discrimination provided by the BGO calorimeter.

The selection relies on the differences in the transverse spread of the shower produced and in the ratio between the energy deposited in the last layer and the total energy released in the calorimeter. Applying this method a proton rejection efficiency of 99.99% has been obtained keeping the electron efficiency at 90%, with a discrimination factor of 10^5-10^6 [7].

The CRE spectrum, shown in figure 2, exhibits a spectral hardening at ~50 GeV in agreement with AMS-02 [8] and Fermi-LAT[9] and provides a first direct and clear detection of a break at about 0.9 TeV, confirming with high precision the feature previously observed by H.E.S.S. [10]. The measured CRE spectrum in the 55 GeV - 2.63 TeV energy range is properly fitted using a smoothly broken power law model, instead of a unique power law. Hence, the spectral index is found to change from ~3.1 to ~3.9 at the ~0.9 TeV spectral break. This study will be pursued including the updated spectrum with larger statistics as well as a more powerful background rejection via the implementation of machine learning algorithms [11], thus extending the analysis to higher energies.

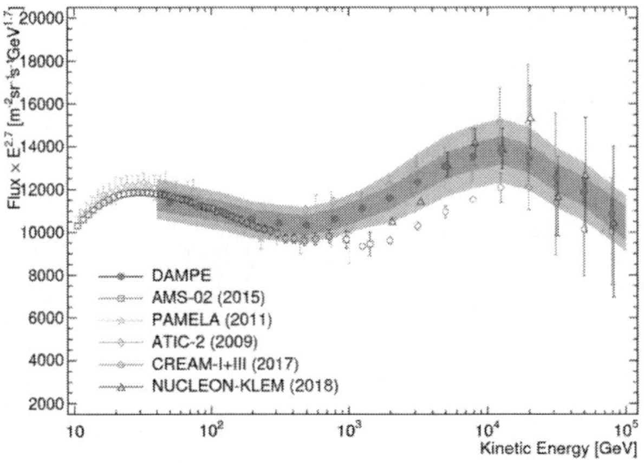

Figure 3: DAMPE proton spectrum (red dots) from 40 GeV to 100 TeV, compared with measurements from other experiments [13–15, 19, 20]. The red error bars refer to the statistical uncertainties. The inner band shows the estimated systematic uncertainties, the outer band refers to the total systematic uncertainties including also those from the hadronic models. The error bars of PAMELA and AMS-02 data include both statistical and systematic uncertainties added in quadrature, while for ATIC, CREAM, and NUCLEON data only statistical uncertainties are shown [12].

3.2 Proton spectrum

The DAMPE detector measured the proton spectrum in the 40 GeV - 100 TeV energy range using its first 30 months of data [12]. The background rejection was performed by using the previously described calorimetric analysis to reject the background contribution due to electrons and positrons, and thanks to the charge selection performed by the PSD, helium and heavier nuclei were discarded. Moreover, a Monte Carlo (MC) based correction that estimated the remaining fraction of helium, after background rejection, was applied.

The obtained proton spectrum (see figure 3) is characterized by a hardening at few hundreds of GeV, previously observed by other experiments [13–18], and revealed a softening at ~14 TeV with a significance of 4.7σ. These spectral breaks show that above the hardening the spectrum is more accurately described by a smoothly broken power law model and from the fit the spectral index changes from ~2.60 to ~2.85 at ~13.6 TeV [12].

3.3 Helium spectrum

The DAMPE helium spectrum, reported in figure 4, was measured in the 70 GeV -80 TeV energy range using 54 months of data. It is characterized by a behaviour similar to the proton spectrum. In fact, it shows an hardening at ~TeV, confirming the previous results observed by other experiments [22–26], and at ~34 TeV it has a softening with high significance, 4.3 σ [21]. For the selection of the He data, the charge measured with the PSD was used. Moreover, also the information from the first plane of the STK tracker was included in the analysis to improve the proton background rejection.

The comparison between the proton flux and helium flux data offers the possibility to get new insights into the acceleration and propagation mechanisms; in particular, the different position of the softening in the proton and He spectra suggests that this feature could be rigidity-dependent, albeit, due to the measurement uncertainties, the hypothesis of a mass-dependent softening cannot be excluded.

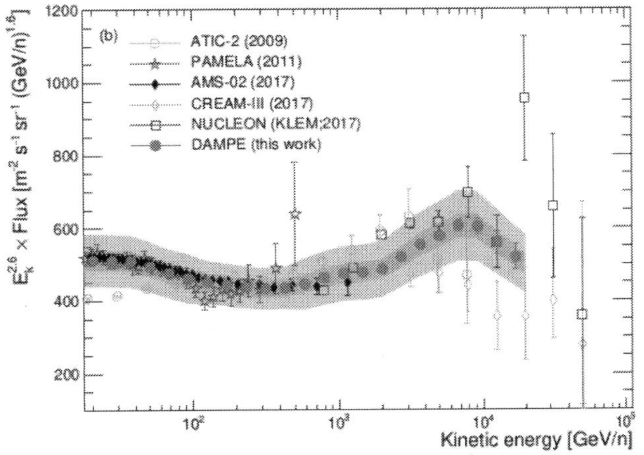

Figure 4: DAMPE He spectrum in kinetic energy per nucleon compared with measurements from other experiments [22, 23, 25, 26]. The red error bars refer to the statistical uncertainties. The inner band shows the estimated systematic uncertainties, the outer band refers to the total systematic uncertainties including also those from the hadronic models [21].

3.4 p+He spectrum

The DAMPE experiment measured also the p+He energy spectrum up to hundreds of TeV [27]. Thanks to the much lower fluxes of the other light nuclei, the sample was selected with larger efficiency and larger purity than the analyses with proton or helium alone. Moreover, looser analysis cuts have been applied allowing to have very high statistics and to cover a more extended energy range. This enables to compare the high energy part of the obtained spectrum with the measurements performed with ground-based experiments. In figure 5 the spectrum obtained with a preliminary analysis is shown. It ranges up to ~150 TeV and is characterized by a hardening at ~600 GeV and a softening at ~25 TeV [27]. This analysis

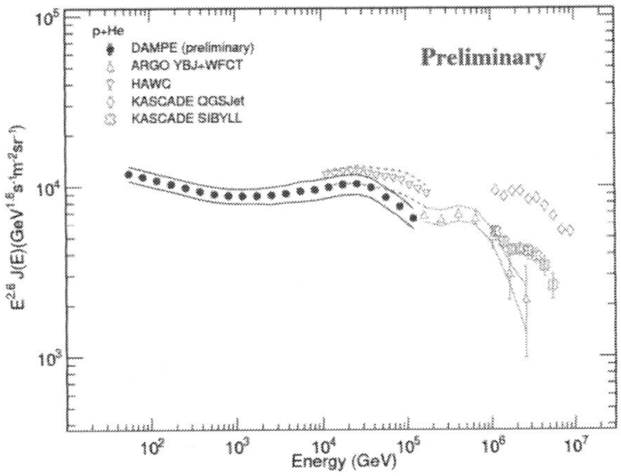

Figure 5: DAMPE p + He spectrum (red dots) weighted by $E^{2.6}$ [27], compared with indirect measurements from other experiments [28–30].

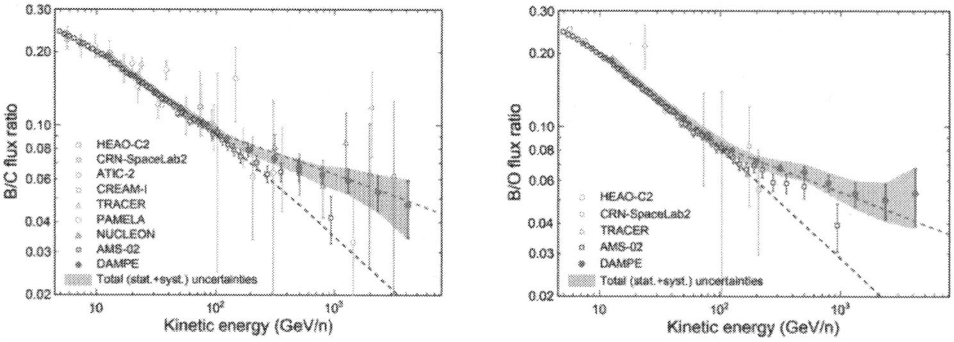

Figure 6: The DAMPE boron to carbon (left) and boron to oxygen (right) flux ratios as a function of kinetic energy per nucleon [31], compared with the measurements from other experiments. The red error bars refer to the statistical uncertainties. The shaded band shows the total systematic uncertainties including those from the hadronic models.

is now being finalized in order to extend the energy range to higher energies and include a complete evaluation of the systematic uncertainty.

3.5 Heavier nuclei

There are various ongoing analyses on heavier nuclei that are studying the secondaries Li and Be, the independent spectra of B, C and O, as well as the boron to carbon (B/C) and boron to oxygen (B/O) flux ratios, and heavier nuclei such as Ne, Mg, Si, up to Fe.

Recent results [31] have been obtained for the secondary-to-primary ratios B/C and B/O, which are of particular interest since boron nuclei are thought to be mainly of secondary origin and to be produced from the fragmentation of heavier primary nuclei, such as carbon and oxygen. Therefore, the measurement of these flux ratios provide information about the CR acceleration and propagation in the Galaxy. The B/C and B/O flux ratios have been measured in the range between 10 GeV/n and 5.6 TeV/n using six years of data. In both cases, a spectral hardening has been observed at $\sim$100 GeV/n as it is shown in figure 6.

3.6 Gamma-ray astronomy and dark matter searches

Gamma-ray analysis was performed by the DAMPE experiment with high precision thanks to its excellent energy and angular resolutions, corresponding to $\sim$1% and $\sim$0.1° at 100 GeV, respectively. The photon events selection was performed using the information from the PSD in order to veto the charged particle events. Moreover, the topology of the shower in the calorimeter was studied to further suppress the proton background component and the STK signal from the first plane was used to reject the surviving electron events.

An analysis for the detection of high energy gamma-ray sources was performed using 5 years of data [32]. A sky map (see figure 7) was produced and it is characterized by 222 identified γ-ray sources which were also associated to the Fermi catalog (4FGL) [33], leading to the identification of several active galactic nuclei and pulsars.

Moreover, the DAMPE gamma-ray studies include the search for sharp line-like features in the photon spectrum [34], since peaked emission lines, if detected, are unlike to be produced by known astrophysical processes and according to some DM models they might be originated from a process involving a DM particle. A search for γ-ray lines was performed

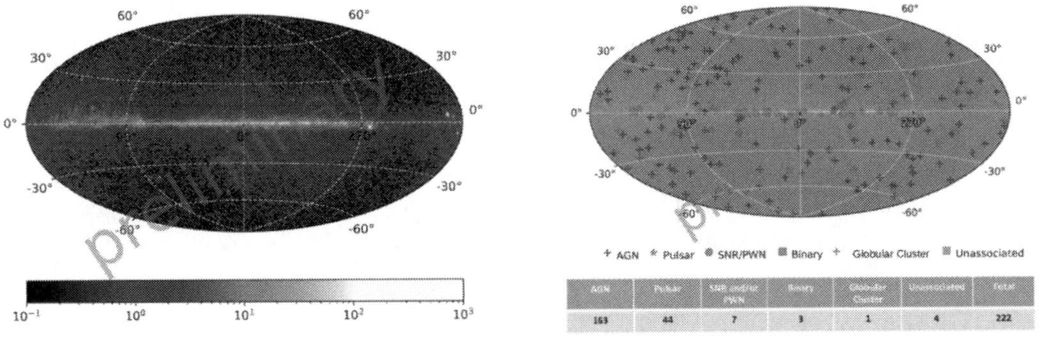

Figure 7: Preliminary sky map (a) and Gamma-ray sources (b) in association with 4FGL observed by DAMPE [32].

using 5 years of data in the 10 - 300 GeV energy range. Since no line signals were detected, a 95% of confidence level constraint was put on the DM decay lifetime and annihilation cross section, as shown in figure 8.

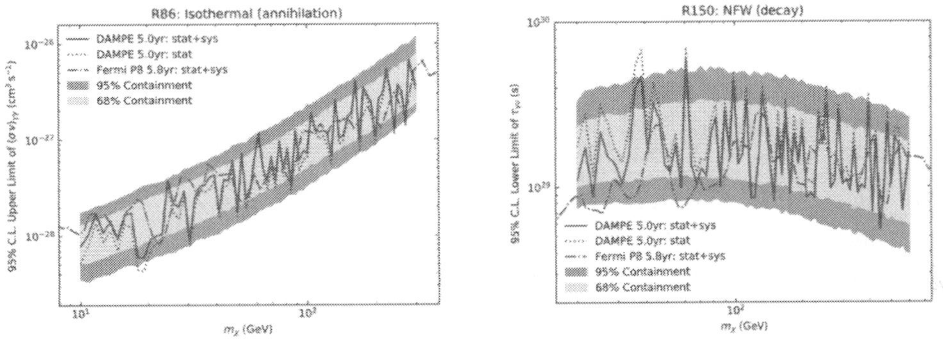

Figure 8: Limits from gamma line searches by DAMPE. A 95% confidence level constraints for different DM models. In the left panel, upper limits on the velocity-averaged annihilation cross-section for DM annihilation are shown. In the right panel the lower limits on the DM decay lifetime are illustrated. The yellow and green bands refer to the 68% and 95% containment respectively [34].

4 Conclusions

The DAMPE detector is operating since December 2015 and in these years of data taking it obtained several results. The electron-positron spectrum was observed over a wide energy range up to 4.6 TeV, providing a first direct and clear detection of a break at about 0.9 TeV. Moreover, the proton and He spectra have been measured up to 100 TeV and up to 80 TeV respectively. In both spectra a hardening at hundreds of GeV/n was confirmed, and for the first time DAMPE revealed a spectral softening at ~14 TeV for protons and at ~34 TeV for He. A preliminary p + He combined spectrum was also been measured in an energy range that extends to higher energies with respect to the independent spectra with the single elements, allowing for a comparison with ground-based experiments results. Other ongoing analyses are focusing on the measurement of primary and secondary CR spectra and recently a study on the B/C and B/O flux ratios was published, showing a hardening at ~100 GeV/n for both flux ratios giving new insight into acceleration and propagation mechanisms of CRs in our Galaxy. In addition, using the first five years of data, DAMPE produced a gamma-ray sky map with the identification of 222 sources, also setting upper limits on the cross section of DM annihilation from gamma line searches.

References

[1] J. Chang et al., Astropart. Phys. (95), 6-24 (2017)
[2] Y. Yu et al., Astropart. Phys. (94), 1-10 (2017)
[3] P. Azzarello et al., Nucl. Instrum. Methods Phys. Res. Sect. A (831), 378 (2016)
[4] Z. Zhang et al., Nucl. Instrum. Methods Phys. Res. Sect. A (836), 98 (2016)
[5] Y-Y. Huang et al., Res. Astrophys. (20), 153 (2020)
[6] G. Ambrosi et al., Astropart. Phys. (106), 18-34 (2019)
[7] G. Ambrosi et al., Nature (552), 63-66 (2017)
[8] M. Aguilar et al., Phys. Rev. Lett.(113), 121 102 (2014)
[9] S. Abdollahi et al., Phys. Rev. D (95), 082007 (2017)
[10] F. Aharonian et al., Phys. Rev. Lett. (101), 261 104 (2008)
[11] D. Droz et al., JINST (16), P07036 (2021)
[12] Q. An et al., Science Advances (5), eaax3793 (2019)
[13] A. D. Panov, et al., Bull. Russ. Acad. of Sci.: Phys. (73), 564-567 (2009)
[14] O. Adriani et al., Science (332), 69 (2011)
[15] M. Aguilar et al., Phys. Rev. Lett. (114), 171103 (2015)
[16] O. Adriani et al., Phys. Rev. Lett. (122), 181102 (2019)
[17] Y. S. Yoon et al., Astrophys. J. (839), 5 (2017)
[18] E. Atkin et al., Astropart. Phys. (90), 69 (2017)
[19] M. Aguilar et al., Phys. Rev. Lett. (119), 251101 (2017)
[20] E. Atkin et al., JETP Lett. (108), 5-12 (2018)
[21] F. Alemanno et al., Phys. Rev. Lett. (126), 201102 (2021)
[22] O. Adriani et al., Science (332), 69 (2011)
[23] A. D. Panov et al., Bull. Russ. Acad. Sci.: Phys. (73), 564 (2009)
[24] Y. S. Yoon et al., Astrophys. J. (839), 5 (2017)
[25] M. Aguilar et al., Phys. Rev. Lett. (115), 211101 (2015)
[26] E. Atkin et al., J. Cosmol. Astropart. Phys. (07), 020 (2017)
[27] F. Alemanno et al., PoS (ICRC2021), 117 (2021)
[28] A. Albert et al., Phys. Rev. D (105), 063021 (2022)
[29] B. Bartoli et al. Phys. Rev. D (92), 092005 (2015)
[30] K. H. Kampert et al. Acta Phys. Pol. B (35), 1799 (2004)
[31] F. Alemanno et al., Science Bulletin (67), 2162-2166 (2022)
[32] K.-K. Duan et al., PoS (ICRC2021), 631(2021)
[33] S. Abdollahi et al.,Astrophy.J.Supp. (247), 33 (2020)
[34] F. Alemanno et al., Science Bulletin (67), 679-684 (2022)

EPJ Web of Conferences **280**, 01002 (2023) https://doi.org/10.1051/epjconf/202328001002
RICAP-22

Ultra-high energy cosmic rays with the Pierre Auger Observatory

Lorenzo Perrone[1,*] for the Pierre Auger Collaboration[2,**]

[1]Universtità del Salento and INFN Lecce (Italy)
[2]Observatorio Pierre Auger, Av. San Martín Norte 304, 5613 Malargüe, Argentina
Full author list: http://www.auger.org/archive/authors_2022_09.html

Abstract.
In the era of multi-messenger astronomy, ultra-high energy cosmic rays offer the unique opportunity to investigate the nature of astrophysical sources and of particle interactions in an energy range far beyond that covered by current particle accelerators. The Pierre Auger Observatory, the world's largest cosmic ray detector, combines in a hybrid design the information from fluorescence telescopes, observing the longitudinal profile of extensive air showers, with a surface array, measuring the lateral distributions of secondary particles at the ground. A review of selected results is presented, focusing on the measurements of energy spectrum and chemical composition and on the search for neutral primary particles. The future prospects will also be discussed in light of the extensive upgrade program being now implemented to further improve the Observatory potential.

1 Introduction

After more than half a century of observations, many of the intriguing mysteries related to the nature and origin of ultra-high energy cosmic rays (UHECRs) remain to be unveiled, though the picture we are reconstructing has now a much higher resolution based on measurements with unprecedented precision. A significant contribution to this field comes from the Pierre Auger Observatory [1] located in the Province of Mendoza, Argentina. The Observatory consists of a surface detector (SD) array of about 3000 km^2 surrounded by 27 air fluorescence detector (FD) telescopes grouped in four sites, providing a powerful instrument for extensive air shower detection. The SD comprises 1600 water-Cherenkov detectors separated by 1500 m (SD-1500) in a triangular grid, plus a smaller nested array of 61 additional detectors spaced by 750 m (SD-750) covering an area of 23.5 km^2. The FD provides a nearly calorimetric estimate of the primary energy, almost independent of the assumptions on hadronic interaction models at the highest energies. The operation of the FD is however restricted to a duty cycle of about 15% because it can operate during moonless nights and it is limited by the atmospheric conditions while the SD measurements are made for about the 100% of the time. The hybrid paradigm relies on the fact that it is possible to calibrate the SD signal using the events simultaneously recorded by the FD, the so-called hybrid events, avoiding to a large extent the use of Monte Carlo simulations and allowing us to achieve good control of the systematic uncertainties in the energy scale [2].

*e-mail: lorenzo.perrone@le.infn.it
**e-mail: spokespersons@auger.org

2 Energy spectrum and composition

The nature and properties of UHECRs and the imprint of propagation effects are embedded in their energy distribution observed at the ground. For this reason, the study of the features characterizing the shape of the energy spectrum has always been among the main goals of the instruments built in this field. The Pierre Auger Observatory can profit from its hybrid design combining a huge accumulated exposure of about 80,000 km^2 sr yr from the SD (SD-1500 + SD-750) with the precise and almost-model-independent energy measurement provided by the FD. The left panel of Fig. 1 shows the energy spectrum for energies above 2.5×10^{18} eV and zenith angles below 60°, derived with more than 200,000 events recorded with the SD-1500 [3]. The SD energy resolution as a function of energy, measured from data, is shown on the right panel in comparison with FD. The systematic uncertainty on the energy scale provided by the FD is 14%, substantially flat in energy and dominated by the calibration of the instrument [2]. The flattening of the spectrum near 5×10^{18} eV (the so-called *ankle*)

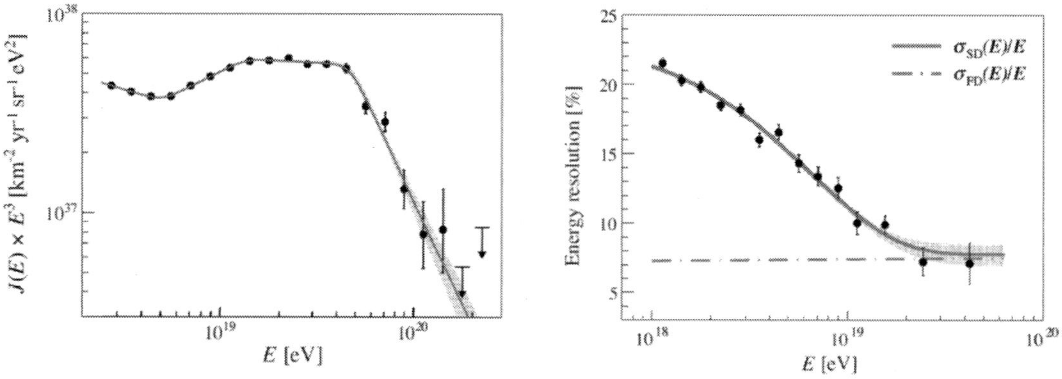

Figure 1. Left: energy spectrum scaled by E^3 and fitted with a broken power-law. The shaded band indicates the statistical uncertainty of the fit. Right: resolution of the SD with statistical uncertainties as a function of energy. The FD resolution is also shown for reference. See [3] for the details about this measurement.

and the subsequent steepening at 5×10^{19} eV are confirmed while a new feature that marks a change from a spectral index of 2.5 to 3.1 named *instep* has been identified in the region above 10^{19} eV. More recently, an extension of the spectrum measurement towards the lower energies has been presented [4] using the SD-750. An inflection of the spectrum starting at about 10^{17} eV is reported, confirming the presence of the so-called *second knee* feature. The combined spectrum SD-1500 + SD-750 is show in Fig. 2, leftmost panel. The energy at which the integral spectrum drops by a factor of two with respect to a power-law extrapolation from lower energies is at about 2×10^{19} eV, which is in tension with that expected for uniformly distributed sources of protons, as also indicated by the results from our composition measurements. Indeed, the average atmospheric depth at shower maximum $\langle X_{\max} \rangle$ is directly proportional to the logarithmic mass of the primary particle, while its variance is a convolution of the intrinsic shower-to-shower fluctuations and of the dispersion of masses in the primary beam. The first two moments of the depth of shower maximum as measured at the Pierre Auger Observatory are shown as a function of energy in the central and rightmost panels of Fig. 2. The observed change of growth rate of $\langle X_{\max} \rangle$ at E_c of about 2×10^{18} eV suggests a non-constant composition that could be interpreted, when comparing with predic-

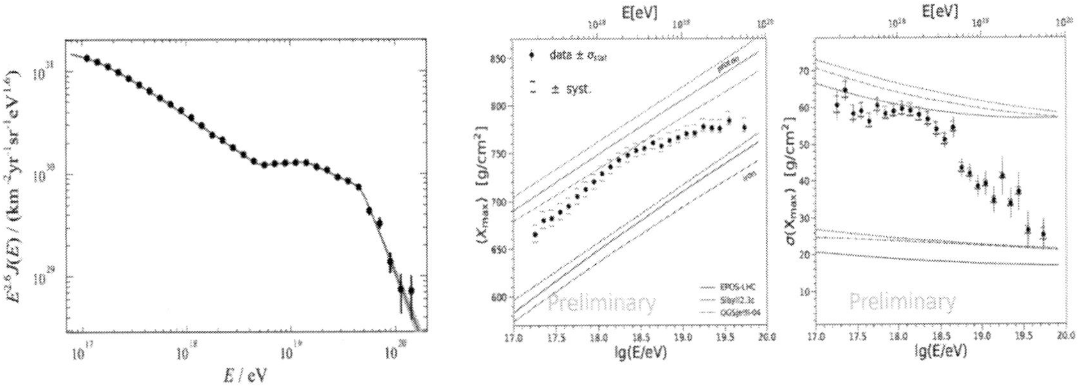

Figure 2. Left: SD energy spectrum after combining the individual measurements by the SD-750 and the SD-1500 scaled by $E^{2.6}$ [4]. Right: the first two moments of the $X_{\max}$ distributions measured by the FD [5].

tions of hadronic models tuned to the most recent LHC results, as an evolution from lighter to heavier elements at the highest energies. The measurement of the standard deviation further supports this result because large values of $\sigma(X_{\max})$ below E_c can be explained as coming from either light or mixed primaries, whereas the subsequent decrease would corresponds to an intermediate to heavy composition.

The measurements of the energy spectrum and mass composition can be combined together with the aim of probing different astrophysical scenarios in terms of source properties, spectral indices at the source, different primary masses and propagation effects [7, 8]. In this context, the flux suppression could be interpreted as mainly due to source exhaustion, rather than to propagation effects. The *instep* would instead reflect the interplay between the flux contributions of the helium and the CNO group injected at the source, shaped by photodisintegration during propagation. In Fig. 3, the contribution of the different mass groups to the energy spectrum and the relative abundances at the top of the atmosphere are shown. The rather large uncertainty on the predicted total fluxes (brown band) is due to combined effects of energy scale and $X_{\max}$ uncertainty. The impact of magnetic fields is not taken into account in these models.

The average composition of UHECRs also appears to be heavier at low than at high Galactic latitudes, the cut being put after a pre-scan at Galactic latitute of 30° [6]. The post-trial significance of this result is currently 4.4 σ. The Galactic magnetic field could be responsible of the mass dependence of the anisotropy, but this effect could be due to the Galactic magnetic field differently affecting different primary masses.

A joint collaboration between the Pierre Auger Observatory and the Telescope Array (TA), a hybrid experiment operating since 2008 in the Northern hemisphere and covering an area of 700 km^2, has started to compare and mutually complement the findings of both experiments. The results on the flux of UHECRs are in agreement within systematic uncertainties up to about 3×10^{19} eV. The difference at higher energies is still under discussion [9]. As for the mass composition, a detailed comparison taking into account the detector effects shows that the TA result is compatible with the mixed composition measured by the Pierre Auger Observatory at least up to 10 EeV [10]. The lack of statistics at the highest energies does not allow TA to distinguish between a mixed, Auger-like composition and a proton one.

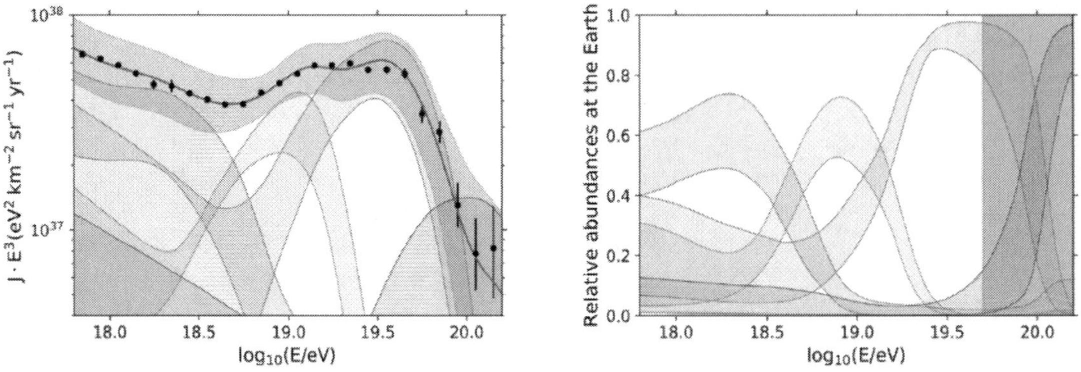

Figure 3. Left: the total energy spectrum (brown) and contribution of various mass group (red: $A = 1$, grey: $2 \leq A \leq 4$, green: $5 \leq A \leq 22$, cyan: $23 \leq A \leq 38$, blue: $A \geq 39$) at Earth predicted by a model [7, 8] fitted to Auger data (black). Right: the relative abundances at the top of atmosphere. The shaded areas indicate the combined effect of systematic uncertainties on E and X_{max}.

3 Anisotropy

The search for anisotropies in the arrival directions of UHECRs is one of the main objectives of the Pierre Auger Observatory's scientific case and a fundamental step towards a coherent interpretation of the observations. This study can be performed at lower energies on large angular scales where the cumulative flux from different sources could possibly be observed despite magnetic deflections. A large-scale dipolar anisotropy in the arrival direction distribution of events recorded by the Pierre Auger Observatory was first observed in [11]. While for energies below 8×10^{18} eV the result is consistent with isotropy, above this energy the first-harmonic modulation in right ascension implies an equatorial dipole amplitude of about 6% with a probability to arise by chance from an isotropic distribution of about 10^{-9}. The sky map in equatorial coordinates is shown in the left panel of Fig. 4. With more than 15 years of data, including events up to 80°, we could exploit the enormous exposure of 120,000 km^2 sr yr. The dipole points $\sim 125°$ away from the Galactic Centre direction, thus suggesting an extra-galactic nature of the UHECRs above this energy. A Galactic origin would result in a much stronger dipole, in a direction within a few tens of degrees of the Galactic Centre.

At energies below 8 EeV, no significant anisotropies were observed, so that relevant upper bounds on the equatorial dipole amplitudes were set down to energies of 0.03 EeV and the right ascension phase of the flux modulations below about 1 EeV is not far from the right ascension of the Galactic Centre. At higher energies, intermediate scale anisotropies can be searched for, as Galactic and extra-galactic deflections of UHECR are expected to be small enough for point sources to be visible as warm/hot spots. The largest available set of events at such energies was obtained thanks to an integrated exposure of 122,000 km^2 sr yr. Some indications of anisotropies at these angular scales, of the order of 25°, have been observed at energies above 41 EeV (right panel of Fig. 4), with the main excess lying around the direction towards the nearby galaxy Centaurus A [12, 13] although other potential sources are located in the same region of the sky, such as the starburst galaxies NGC4945 and M83.

Evidence of a deviation from isotropy has been also found by correlating the arrival directions of UHECRs with the direction of starburst galaxies and active galactic nuclei (AGN) above $\sim 40 \times 10^{18}$ eV, at 4.2 σ and 3.3 σ significance level respectively [12]. This result has been derived accounting for the attenuation of the UHECR nuclei, assuming a mixed mass

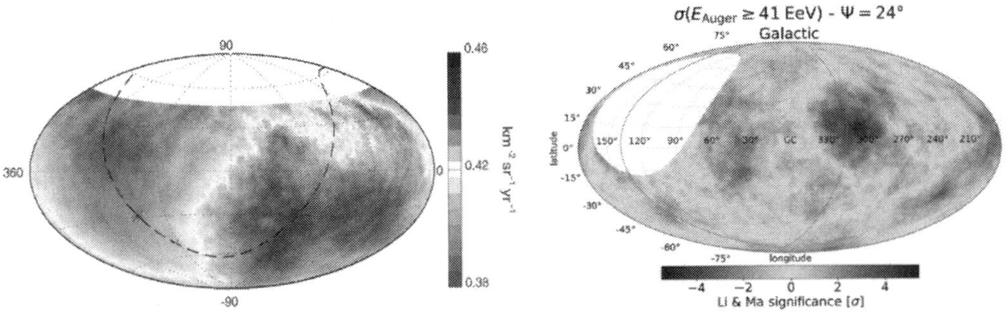

Figure 4. Left: Sky map in equatorial coordinates, using a Hammer projection, showing the cosmic-ray flux above 8 EeV smoothed with a 45° top-hat function. The Galactic Centre is marked with an asterisk; the galactic plane is shown by a dashed line [11]. Right: Local Li–Ma significance map at energies above 41 EeV and within a top-hat search angle of $\Psi = 24°$ in Galactic coordinates [12].

composition as inferred from lower energy observations. The observed correlation does not allow us to disentangle different classes of objects yet; it could furthermore be altered by the effects of Galactic and extra-Galactic magnetic fields.

The comparison with the results obtained by TA allows us to explore possible deviations from isotropy over the full sky. The search for large scale anisotropies over the full sky confirmed the Auger results described above, with reduced systematic uncertainties [14]. Studies on intermediate scale anisotropies found excesses lying along the Supergalactic Plane; about 12% of the events show a correlation with the position of nearby starburst galaxies on a 15° angular scale, with a post-trial significance of 4.2 σ [15].

4 Multi-messengers

UHECRs are mostly protons and other atomic nuclei; a fraction of photons would be an ideal probe for fundamental questions in particle physics and cosmology, such as the nature of dark matter and the possibility of Lorentz invariance violation. Indeed, the decay of super-heavy DM particles could result in large fluxes of UHE Standard Model particles [16], with photons and neutrinos dominating the final state. Also, UHE photons could result from the decay of hadrons produced when protons with ~ 10 the final photon energy interact with microwave/IR/visible/UV background photons, either within the source environment or during their propagation through the intergalactic medium. In the Standard Model, such UHE photons are expected to be quickly absorbed by background photons, but LIV phenomena could inhibit such interactions, resulting in a significantly higher fraction of photons in the UHECR flux. Only UHECR observatories have capabilities to detect photons at these energies, through the extensive air showers they produce in the atmosphere. In this context, the Pierre Auger Observatory has the highest sensitivity to photon primaries. Due to the lower multiplicity of electromagnetic interactions (compared to hadronic), photons induced showers develop deeper in the atmosphere and have less muons thus yielding a larger depth at shower maximum, $X_{\max}$, and a steeper lateral distribution function. These distinctive features are also reflected in a longer rise time for the signal shape observed with SD. Since the start of data collection, various searches for ultra-high-energy with energy above 10^{17} eV photons have been performed, either for a diffuse flux, for point sources or for photons associated with transient events such as gravitational wave events. A very recent review of current results in this context can be found here [18] and details about the detection techniques for specific

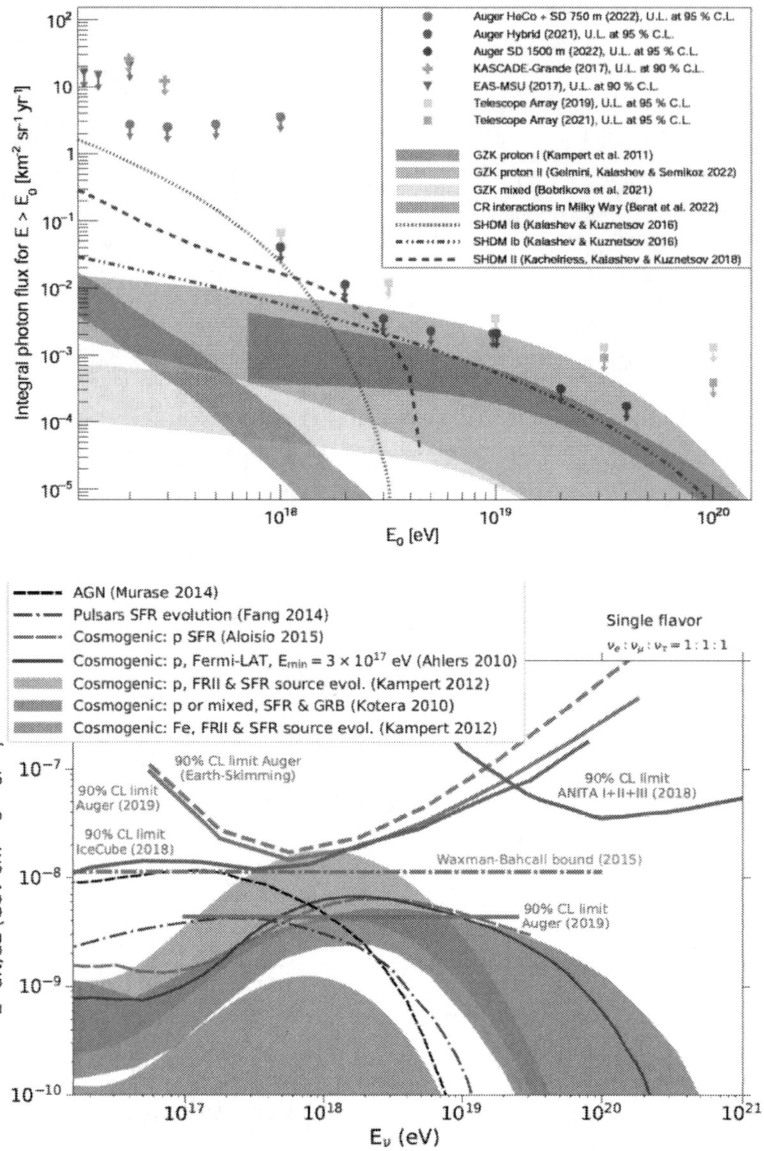

Figure 5. Upper limits on the cosmogenic photons (top) and neutrinos (bottom) fluxes compared to limits from other experiments and to model predictions [19, 20] and references therein.

photon searches have been presented at this conference [17]. As an example, the current status of the search for a diffuse flux of photons is shown in Fig. 5, top panel. Upper limits set by the fluorescence detector operating in hybrid mode play an important role at the lowest and intermediate energies, up the EeV range, those set by the surface detector dominate at the highest energies, above 10^{19} eV.

Several model predictions in the context of dark matter scenarions are severely constrained by the Auger results and also the expectation of photons originating from the interaction of UHECRs with the interstallar medium start to be in tension with a scenario of a proton-dominated chemical composition. Thus, the search of photons can provide strong bounds on fundamental physics and, at the same time, valuable indications on UHECRs mass composition.

While photons have a limited horizon due to their interaction with the interstellar radiation field (few Mpc at 10^{18} eV), neutrinos can travel substantially unabsorbed over cosmological distances. Their fluxes depend on the properties of the sources and on the composition of the primary beam. Neutrinos can be searched for by selecting horizontal showers with a large electromagnetic component close to the detector. This provides a clear identifier of a neutrino-induced shower as, for hadronic cascades, the electromagnetic component would be almost completely absorbed and only muons could be detected in the SD. Two main categories of events are considered: Earth-skimming, induced by ν_τ travelling upwards from below the Earth crust in directions just below the horizon and producing a τ lepton which can then generate a shower above the SD, and downward-going events due to neutrinos of any flavour. Fig. 5, bottom panel, shows the most recent upper limits on cosmogenic neutrinos [20], along with model predictions and limits from other experiments.

The FD detector is also sensitive to upward-going events and can be used to search for upward-going showers without limitation on the zenith angle of the emerging induced showers. The first results from this study, triggered in the context of what reported by the ANITA Collaboration in the energy range above 10^{17} eV, have been reported here [21].
Given the very good angular resolution of the Observatory (less than 0.5° for zenith > 60°) specific follow-up searches were performed with the SD detector looking for point-like sources in spacial and time coincidence with gravitational waves events: upper limits on the total energy per source radiated in neutrinos [22] and on the photon fluence [23] have been recently set.

5 Dissemination

The Pierre Auger Collaboration is committed to the public release of their data for the purpose of re-use by a wide community including professional scientists, in educational and outreach initiatives, and by citizen scientists. The current policy allows a release of 10% of the cosmic-ray data and 100% of atmospheric data in close-to-raw (JSON) format with summary CSV file containing the reconstructed parameters. In addition to hosting data for downloading, various tools for analysis (with Python notebook examples), event visualisation and a specific outreach section are available on the dedicated website https://opendata.auger.org/. A comprehensive description of this activity is reported here [24].

6 Perspectives

The Pierre Auger Observatory is currently being upgraded within a framework named AugerPrime [25]. The scientific case to be addressed with the upgraded detector includes a better understanding of the origin of flux suppression at the highest energies, the search for a flux contribution of protons up to the highest energies at a level of 10% and the study of extensive air showers and hadronic physics at $\sqrt{s} \sim 70$ TeV. This will require an improvement in the detector ability to identify the composition of primary cosmic rays on an event-by-event basis, which can be achieved in practice by disentangling the muon and electromagnetic components of extensive air showers. The upgrade program includes the deployment of an additional 4m^2 scintillator layer on top of each water-Cherenkov detector [26], an additional small-area photomultiplier tube [27], and an additional radio antenna to observe the radio emission from the electromagnetic component of the shower [28]. The current electronics will also be replaced by faster and newer boards, with sampling frequency of the signals increased by a factor 3 [29]. Finally, the SD-750 and the SD-433 will be equipped with underground scintillators to provide a direct measurement of the muonic component of the extensive air showers. [30]. All the additional scintillator layers have been installed and about 40% of the array is already equipped with the new electronics. The goal is to complete the deployment by mid-2023. The results of the commissioning data analysis show good uniformity and long-term performance. The Collaboration aims at running the upgraded Observatory until 2030. This would yield an additional vertical exposure of 40,000 km^2 sr yr. Furthermore, a campaign is planned to re-analyze the Phase I dataset by using machine learning techniques, taking at the same time advantage of the AugerPrime Phase II data.

References

[1] The Pierre Auger Collaboration, Nucl. Instrum. Meth. A **798**, 172 (2015).
[2] B. Dawson (Pierre Auger Coll.), PoS(ICRC2019)231.
[3] The Pierre Auger Collaboration, Phys. Rev. D **102** (2020) 062005.
[4] The Pierre Auger Collaboration, Eur. Phys. J. C **81** (2021) 966.
[5] A. Yushkov (Pierre Auger Coll.), PoS(ICRC2019)482.
[6] E. Mayotte (Pierre Auger Coll.), PoS(ICRC2021)321.
[7] E. Guido (Pierre Auger Coll.), PoS(ICRC2021)311.
[8] A. Condorelli (Pierre Auger Coll.), at this conference.
[9] Y. Tsunesada (Pierre Auger and Telescope Array Coll.), PoS(ICRC2021)337.
[10] A. Yushkov (Pierre Auger and Telescope Array Coll.), EPJ Web of Conf. **210** (2018) 010009.
[11] The Pierre Auger Collaboration, Science **357** (2017) 1266-1270.
[12] The Pierre Auger Collaboration,The Astrophysical Journal **935** (2022) 170.
[13] L. Caccianiga (Pierre Auger Coll.) at this Conference.
[14] P. Tinyakov (Pierre Auger and Telescope Array Coll.), PoS(ICRC2021)375.
[15] A. di Matteo (Pierre Auger and Telescope Array Coll.) at this Conference.
[16] R. Aloisio (Pierre auger Coll.), at this Conference.
[17] E. Guido (Pierre Auger Coll.), at this Conference.
[18] The Pierre Auger Collaboration, Universe **8** (2022) 579.
[19] The Pierre Auger Collaboration, Astrop.J. **933** (2022) 125.
[20] The Pierre Auger Collaboration, JCAP **10** (2019) 022.
[21] M. Mastrodicasa (Pierre Auger Coll.), PoS(ICRC2021)1140.
[22] M. Schimp (Pierre Auger Coll.), PoS(ICRC2021)968.
[23] P. Ruehl (Pierre Auger Coll.), PoS(ICRC2021)973.
[24] V. Scherini (Pierre Auger Coll.), PoS(ICRC2021)1386.
[25] The Pierre Auger Collaboration, The AugerPrime Design Report, arXiv:1604.03637.
[26] G. Cataldi (Pierre Auger Coll.), PoS(ICRC2021)251.
[27] A. Castellina (Pierre Auger Coll.), PoS(ICRC2017)397.
[28] F. Schlüter (Pierre Auger Coll.), PoS(ICRC2021)262.
[29] G. Marsella (Pierre Auger Coll.), PoS(ICRC2021)230.
[30] A. M. Botti (Pierre Auger Coll.), PoS(ICRC2021)233.

LHAASO Status and Physics Results

Zhen Cao[1,2,3,*] *on behalf of* LHAASO Collaboration

[1]Key Laboratory of Particle Astrophyics, Institute of High Energy Physics, Chinese Academy of Sciences, 100049 Beijing, China
[2]University of Chinese Academy of Sciences, 100049 Beijing, China
[3]Tianfu Cosmic Ray Research Center, Chengdu, Sichuan, China

Abstract. Status of the LHAASO experiment and its latest results in both γ-ray astronomic observations and charged cosmic ray (CR) measurements are reported in this paper. The discovery of photons around 1 PeV from the Crab, 12 PeVatrons in our galaxy and new sources with emissions above 100 TeV declare the onset of the utral-high-energy γ-ray astronomy. The capability of measurements of spectra of the individual species of CRs, e.g. protons and iron nuclei, starts the new era of the high precision measurement of CRs using the ground based extensive air shower technique. The detection of the highest energy photon at 1.4 PeV from the remote potential source in Cygnus region allows to test of the violation of fundamental physics rules, such as the Lorentz Invariance, and set the highest limits in the tests.

1 Introduction

Very High Energy (VHE) γ-ray astronomy has been making very important progresses with many historical discoveries in understanding the non-thermal universe, such as the discovery of the Crab as the first source in TeV band, Mrk421 as the first extragalactic source associated with the highly variable AGN, very difficult efforts on finding PeVatrons at the center of our galaxy and recent observation of TeV afterglow of GRBs by HESS and MAGIC. In last decode, extensive air shower detection technique has been developed at high altitudes to have sufficient sensitivity for TeV γ-ray sources and making discoveries, such as pulsar halos. However, the lack of observational data above 80 TeV due to the limited sensitivity at higher energies, start to constrain the development of the field. Not only in the γ-ray source search, the charged cosmic ray (CR) measurements and corresponding modeling/theory have also been stuck with low quality data in the very important range in which the knees of spectra of CR species, i.e. proton, helium and iron nuclei may exist. Comparing to the very precise measurements at energy below 100 TeV by many space borne detectors, the precision measurements of CRs in knee range, to release from the entanglement between the composition and interaction in shower development in particular, is more than necessary. To answer the challenges, the Large High Altitude Air Shower Observatory (LHAASO) was designed and started to construct in 2016. In July 2021, the whole complex of detector arrays as a specified instrument for the EAS detection was built completely. The operation with the partially finished arrays started in 2019 and with the full scale in last year are very stable

*e-mail: caozh@ihep.ac.cn

with the duty cycle of 98% and detector failure rate <2%, thus maintains an unprecedented γ-ray detection sensitivity above 20 TeV and highest survey sensitivity of 10 mili-CrabUnit in VHE band. The capability of separation between proton from other CR species and iron from others light up hopes to improve our knowledge about the knee of CR spectrum, thus of the propagation properties of CRs in our galaxy and the mechanism of the knee. Data collected by LHAASO in last two years really tell us the difference in the new territory of the non-thermal universe from what we have learnt in last few decades. A new era, ultra-high-energy (UHE) astronomy, has been initiated. The status of the LHAASO experiment and the physics results in both γ-ray astronomy and CR measurements are reported in this paper.

2 LHAASO Experiment and the Status

LHAASO had been completely built on top of Mt. Haizi (4410 m above sea level) in Sichuan province, China by July 2021[1]. The vertical atmospheric depth is ~600 g/cm^2. Since then, the kilometer-square array of 5216 scintillator counters and 1188 μ-detectors (KM2A) and water Cherenkov detector array (WCDA) gaplessly covering 78,000 m^2 with 3120 detector cells are fully operated. The wide field of view (FoV) Cherenkov telescope array (WFCTA) composing of 18 telescopes with a 5 m^2 light collector for each is also operated in dry seasons for more than 1400 hours per year.

In KM2A, scintillator counters are deployed in a triangle grid with a space of 15 m between counters which is covered by a 5 mm thick lead plate, which converts secondary γ-rays into e^+e^- pairs in the counter, and measures the total deposited energy by all charged particles passing through the scintillator plates. The number of particles is measured according to the charge of single particle measured during the data taking. μ-detectors are also in the triangle grid with a distance of 30 m between detectors, and the total active area is 40,000 m^2. With the overburden of 2.5 m soil. The muon detection has nearly zero backgrounds of e^+s, e^-s or γ-rays in a shower, except for one or two detectors very closed to the shower core. The μ-content with the threshold energy of muons ~0.9 GeV is measured for each shower. The lateral distributions of both electromagnetic particles and muons in a shower are well measured using KM2A. μ-content is a very useful parameter for cutting CR background effectively in γ-ray astronomic observation. In the CR composition analysis, it is a sensitive parameter to the CR composition, as well.

In WCDA, all energy carried by e^+s, e^-s and γ-rays in a shower is deposited in water inside the detector. Cherenkov light generated in water is proportion to the deposited energy and sampled by the PMTs at the center of each cell and 4 m beneath the water surface. Muons in the shower could generate larger signals than electromagnetic particles since they have longer tracks in water. They could generate exceptionally large signals in cells that the muons pass through, particularly for those cells far from shower cores. In γ-ray detection, this can serve as a discriminator to suppress CR background. For showers with energy higher than 10 TeV in CR measurements, the lateral distribution of the energy flux is well measured as a rather smooth function. The extra fluctuation due to μ's is also useful in CR composition analysis.

Cherenkov telescopes are operated at higher threshold energy than those of WCDA and KM2A, so events recorded by the telescopes can find their counterparts either in WCDA or/and in KM2A. With the well reconstructed shower geometry by the arrays, the images recorded by the telescopes can be used to extract a couple of shower parameters. The shower energy can be reconstructed with the total number of photons in the image by knowing exact distance between the telescope and the shower axis. Simultaneously, the angular distance between the centroid of the image and the arrival direction measures the atmospheric depth of the shower maximum as well. The shape of image has a correlation with the shower composition at certain level.

All components of LHAASO are operated smoothly with very stable performance since July 2021 when all the detectors were deployed. KM2A and WCDA the two detector arrays for the γ-ray astronomy with a duty cycle more than 98%. Daily rate is 2×10^8 events above 10 TeV and 3×10^9 events above 300 GeV, respectively.

Up to now, 70 million CR events accumulated by WFCTA and matched with either WCDA or KM2A, thus forming a hybrid data set of CRs above 10 TeV. The shower geometry is reconstructed using the EAS arrays with very high resolutions, i.e. the shower arrival direction and the core location are measured with errors smaller than 0.2° and 3 m, respectively. Further analysis for CR composition and shower energy measurements are under development.

2.1 Scientific Results: Gamma Ray Astronomy Using KM2A and WCDA

As a γ-ray detector, LHAASO covers the very unique energy range from few hundred GeV to few PeV, the highest energy of photons expected from the universe. With a wide acceptance angle, which covers 1/6 of the sky at any moment and operates for 24 hours, LHAASO measures the diffuse γ-rays from the entire northern sky and surveys for sources. Obviously, such a detector is particularly good in monitoring for transient phenomena such as GRB and searching for spatially extended structure of γ-ray emission. Here in the γ-ray astronomy, we have a standard candle, the Crab, to be measured for calibration of the detector at energies it has been well measured by all other detectors in the whole world up to tens of TeV. Of course, as the very unique object, the Crab is discovered over again by LHAASO as a surprising PeVatron. Let us start with the Crab.

2.1.1 The Crab

At least three parameters, i.e. pointing error, angular resolution and photon flux at energies below 100 TeV are well measured in previous experiments by using different EAS detection techniques or in other band. WCDA and KM2A have measured the Crab with very high statistic significance, i.e. 300σ at E>1 TeV and 86σ at E>25 TeV, respectively. Therefore the direction of the Crab as a point-like source is measured with an error of 0.01°, thus the pointing direction of WCDA and KM2A is calibrated within at least 0.01°. The angular distribution of the significance of the Crab measures the resolution of WCDA and KM2A of 0.22° for E>1 TeV, 0.30° for E>25 TeV and 0.15° for E>100 TeV, respectively. The photon flux is measured from 0.5 to 100 TeV by WCDA and KM2A with an overlap at 12 TeV. The agreement between the two sub-arrays within the statistic error has realized a cross checking inside the collaboration and the overall Spectral Energy Distribution (SED) also agreed with other experiments, including experiments using IACT and EAS techniques, as shown in Figure 1. The exciting results from LHAASO is the measurement of the SED in UHE band, i.e. above 100 TeV marked in pink. Firstly it shows a power-law shape SED with the index of -3 up to 2 PeV. Secondly the highest energy photon detected from the Crab reaches 1.1 PeV. This gives a model-independent evidence of existence of an accelerator of either electrons or protons at energies at few PeV in the nebula. The electron accelerator seems more favored because the data could be successfully explained well enough up to 50 TeV even with a simple one-zone model that can produce photons up to few hundred TeV. This strongly implies the existence of so-called extreme accelerator that has an acceleration rate η higher than 15% which is 3 orders of magnitudes larger than the normal shock wave in SNRs. LHAASO data seems to favor a high energy proton component, with an improved fitting quality of the SED

 https://doi.org/10.1051/epjconf/202328001003

from 1 GeV to 2 PeV. In such a scenario, the protons with a spectrum of E^{-2} exponentially cut around tens of PeV are found to be mainly responsible to the photon emission at energies above 500 TeV. More PeV photons will be collected by LHAASO in following years for more conclusive selection between scenarios.

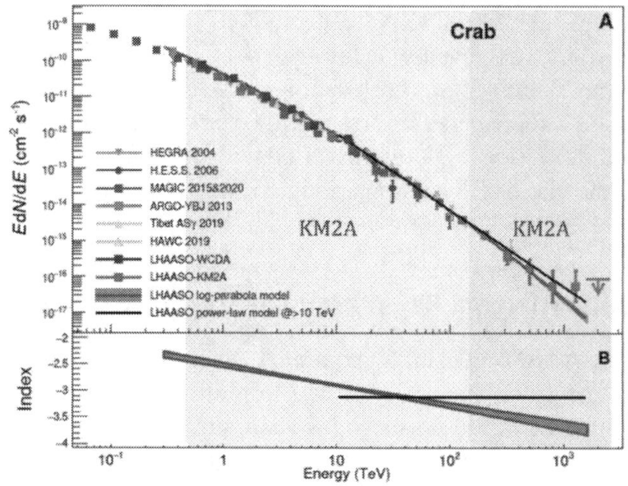

Figure 1. The SED of the Crab Nebula measured by LHAASO-WCDA from 0.5 to 12 TeV and LHAASO-KM2A from 12 TeV to 3 PeV together with other 7 experiments (upper panel). The figure is quoted from reference [2]. The energy range marked in green has been covered by the previous experiments, see the references in [2]. 7 of them are shown here with good agreement with LHAASO measurements. In the lower panel, the spectral index as function of energy is shown with gradual softening and the SED could also be fitted well enough with a power-law in UHE band above 100 TeV with the index around -3.

2.1.2 The Pevatrons

12 new γ-ray sources discovered by LHAASO with emission in UHE band, including the sole identified Crab as a nearly isolated pulsar wind nebula. This reveals the existence of PeVatrons as the central engines boosting particles at energies higher than 1 PeV and emitting the observed photons in interactions with ambient materials. The PeVatrons are lining up along the Galactic plane and 11 out of the 12 have their counterparts found in VHE band i.e. above 100 GeV. Similar to the Crab, the SEDs of the bright ones among the 12 do not show significant cut-off around 100 TeV. This implies that 1. PeVatrons are widely existing in our galaxy and 2. the limit on the particle acceleration capability around 1 PeV for the Galactic accelerators is not real. Amazingly, those discoveries are realized only by the partially finished LHAASO-KM2A (1/2 of designed size) with the observation in first 11 months. The μ-detector array played the key role in γ-ray selection with a CR background rejection power of 10^{-4} at 100 TeV and 10^{-5} above 500 TeV. The large collection area of KM2A enables a collection of 530 UHE photons distributing spatially in 12 clusters with significance greater than 7σ for each. Except for the number of the PeVatrons, which is a surprise to the community that has hunted the PeVatrons for years, the divergence of types of potential counterparts of those UHE γ-ray sources conveys very important information about the origin of the particles that emit the UHE photons. It can be read from Table 1[3] and the references in [3]. No

Table 1. The PeVatrons discovered by LHAASO-KM2A in the operation with the partially finished array in the first 11 months. This is the Version-0 LHAASO UHE γ-ray sources with their name, possible origin, type of the celestial body of the origin, distance, age and spin-down luminosity for pulsars are listed for each source.

LHAASO Source	Possible Origin	Type	Distance (kpc)	Age (kyr)	L_s (erg/s)
LHAASO J0534+2202	PSR J0534+2200	PSR	2.0	1.26	4.5×10^{38}
LHAASO J1825-1326	PSR J1826-1334	PSR	3.1 ± 0.2	21.4	2.8×10^{36}
	PSR J1826-1256	PSR	1.6	14.4	3.6×10^{36}
LHAASO J1839-0545	PSR J1837-0604	PSR	4.8	33.8	2.0×10^{36}
	PSR J1838-0537	PSR	1.3	4.9	6.0×10^{36}
LHAASO J1843-0338	SNR G28.6-0.1	SNR	9.6 ± 0.3	$\lesssim 2$	—
LHAASO J1849-0003	PSR J1849-0001	PSR	7	43.1	9.8×10^{36}
	W43	YMC	5.5	—	—
LHAASO J1908+0621	SNR G40.5-0.5	SNR	3.4	$\sim 10 - 20$	—
	PSR 1907+0602	PSR	2.4	19.5	2.8×10^{36}
	PSR 1907+0631	PSR	3.4	11.3	5.3×10^{35}
LHAASO J1929+1745	PSR J1928+1746	PSR	4.6	82.6	1.6×10^{36}
	PSR J1930+1852	PSR	6.2	2.9	1.2×10^{37}
	SNR G54.1+0.3	SNR	$6.3^{+0.8}_{-0.7}$	$1.8 - 3.3$	—
LHAASO J1956+2845	PSR J1958+2846	PSR	2.0	21.7	3.4×10^{35}
	SNR G66.0-0.0	SNR	2.3 ± 0.2	—	
LHAASO J2018+3651	PSR J2021+3651	PSR	$1.8^{+1.7}_{-1.4}$	17.2	3.4×10^{36}
	Sh 2-104	H II/YMC	$3.3 \pm 0.3/4.0 \pm 0.5$	—	—
LHAASO J2032+4102	Cygnus OB2	YMC	1.40 ± 0.08	—	—
	PSR 2032+4127	PSR	1.40 ± 0.08	201	1.5×10^{35}
	SNR G79.8+1.2	SNR candidate	—	—	—
LHAASO J2108+5157	—	—	—	—	—
LHAASO J2226+6057	SNR G106.3+2.7	SNR	0.8	~ 10	—
	PSR J2229+6114	PSR	0.8	~ 10	2.2×10^{37}

doubt, this initiates a new era of the γ-ray astronomy in the new band of γ-ray energy greater than 100 TeV, the highest energy band for electromagnetic waves in the universe.

2.1.3 New VHE/UHE γ-ray Source Catalog

As mentioned above, one of the PeVatrons does not have its counterpart found in TeV catalog. Actually in lower energy range several new source have been discovered already. LHAASO J0341+5258, J2108+5157 and J0622+3794 are three newly discovered VHE sources have published with detailed analyses about their morphology and SEDs[4–6]. The last has been found a candidate of the pulsar halo. Many more new sources have been discovered in the LHAASO observation in last two years. As discussed in the beginning of this section, LHAASO has capability of finding sources with large spatial extension which is a common feature of the new sources. The new catalog as LHAASO-Ver-1 will be published very soon.

2.2 Scientific Results: Cosmic Ray Measurements Using Hybrid detection Technique

Identification of primary particle of EAS is a very challenging task. Measuring as many independent parameters of showers as possible is a promising way to approach. LHAASO, which is equipped with at least four types of detector sub-arrays, offers the opportunity to identify the primary particle species fore individual detected shower at energies from 100 TeV to 100 PeV. This is a very interesting energy range in which the knees of spectra of proton, helium and iron nuclei may appear. The flux around the knee of the proton spectrum is approximately 5 orders of magnitudes higher than the iron flux around its knee which might be at energies higher than the proton knee by a factor of 26 or 56. Obviously, different configurations of the hybrid detector arrays will be used for the proton and iron spectrum measurements. The experiment is naturally divided into two phases according to the schedule of construction of the detector arrays.

2.2.1 Multi Parameter Measurement of EAS

A typical hybrid shower event with the core inside WCDA-1 recorded in Phase-I is shown in Figure 2. In the right-upper panel, WCDA-1 measures the shower core with a huge dynamic range up to 0.2 million photoelectrons recorded in a cell as indicated by the color scale in the figure. This allows a measurement of the fine structure of the lateral distribution of the energy flux in the core. Here, shower particles have not experienced much scattering in the air so that they represent the fresh population generated not far above the surface, in other words, the hadronic components of the shower before them hit the surface. The energy flux weighted average lateral distance r_i of detector cells $< ER >= \Sigma Q_i r_i / \Sigma Q_i$, where Q_i is the charge measured by the i-th cell, is chosen as a parameter sensitive to the primary species, denoted as p_F after normalized by the preliminarily determined shower energy. The right-lower panel in Figure 2, the detection of the muon lateral distribution is shown as the number of muons in each MD. It is noticed that only handful muons recorded by each MD at distance more than 100 m from the core. Summing over all MDs between 30m and 380 m from the core for the total measured number of muons. It is a rather traditional parameter for species selection in EAS observation due to the relation $ln\mathrm{N}_\mu \sim 0.1 ln\mathrm{A}$, where N_μ is the muon content of the shower induced by a primary particle with the atomic number A. Normalized by using the total muon content in the shower, the parameter P_μ is selected as a sensitive variable for primary species identification. In the left-upper panel of Figure 2, a typical image of the shower is recorded in the camera of WFCTA. It is well known that the shower elongates longer for proton initiated shower than that by heavier nuclei. However, the length of the image depends the distance of the the shower to the telescope, R_p, due to a pure geometrical effect. Corrected by R_p and the preliminarily reconstructed shower energy, the parameter defined as the ratio of the length and the width of the image, denoted as P_C, is found sensitive to the composition of the primary species.

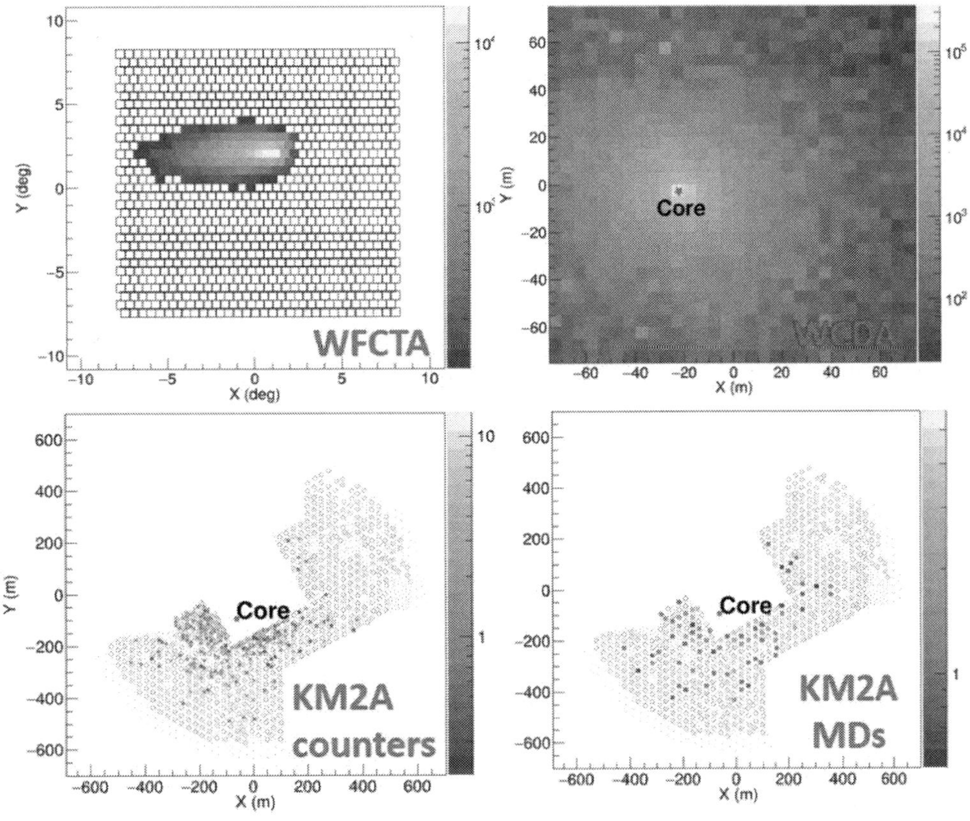

Figure 2. A typical hybrid event recorded in Phase-1.

2.2.2 Measuring the pure proton spectrum

The first phase started with 6 telescopes deployed at the southwest corner of WCDA-1, i.e. the first built pool of 150m×150m for the 900 cells of the LHAASO-WCDA, from 2019-2020 observational season. All telescopes have the elevation angle of 60°, i.e. the zenith angle range from 22° to 38° is covered, so that there is at least 647 g/cm^2 for showers to develop in the air. Before April 2021, the hybrid operation with this configuration, had accumulated more than 70 million events with the cores inside WCDA-1 for 720 hours during which 1/2 of KM2A were operated.

Examing the capability of the parameters individually in the primary particle identification with a simulation toolkit that takes all detector effects into account for all four type of detectors in LHAASO, it seems very challenging if a high purity of the proton sample, e.g. 90%, is the demand, the event selection efficiency is only ~20%. On the other hand, correlations between those parameters are found not negligible. Further efforts to find more independent parameters is still on going. Moreover, for cores falling in KM2A, P_μ and P_C are available. Selection efficiency using P_μ is even improved due to better muon content is measured than those in WCDA. Large statistics provided by the huge KM2A, makes this measurement very useful, particularly serves as cross checking.

2.2.3 Measuring the Fe spectrum

In phase II starting from September 2021, all 18 telescopes are deployed to the sites near the southeast corner of WCDA-1, and form a ring-shaped FoV covering all azimuthal direction between zenith angle range from 37° to 53°. Such a setup is aiming at the detection of CRs at higher energies around the possible knee of iron spectrum, i.e. >10 PeV, so that more atmospheric depth, >751 g/cm^2, allows better developments of showers. To collect sufficient statistics,the entire KM2A needs to be used to achieve the hybrid measurements. The geometrical factor enhancement of a factor of 3 due to the total field-of-view (FoV) of telescopes and a factor of 44 due to the acceptance of the ground array are expected from the re-configuration of the experiment. In the operation for 6 years, a sample of pure iron induced showers with a smaller statistics, roughly estimated to be 20 times smaller than the proton sample, would be expected.

2.2.4 Energy Reconstruction

For pure proton or iron samples, the shower energy reconstruction is rather straightforward by using the total charge measured in the Cherenkov image of a shower at the impact parameter R_p from the telescope. Given the well determined shower geometry by using the EAS arrays, the shower energy uncertainty is dominated by the fluctuation of shower longitudinal development. In case of proton showers, quite symmetric resolution functions are found to be fitted with a Gaussian form with the σ to be 20% around 0.1 PeV and 15% above 1 PeV. The systematic shift of the Gaussian central expectations are less than 2% over the energy range from 0.1 to 10 PeV. The energy response function for showers at impact parameters of 120-130 m from the telescope, resolution and systematic shift as functions of shower energy and a resolution distribution at 1 PeV are shown in Figure 3. Some non-Gaussian tails are found beyond 2σ region due to the deeply developed showers were favored in the proton selection procedure. Further modification of the energy reconstruction by taking into account the EAS parameters is underdevelopment to improve the resolution.

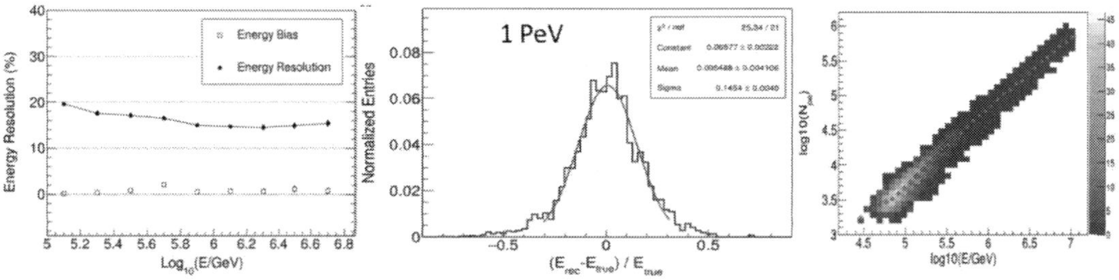

Figure 3. Energy reconstruction for pure proton samples. In the left panel, the shower energy resolution and systematic shift are shown over the energy from 100 Te V to 5 PeV. In the middle panel, the energy resolution function, i.e. the distribution of relative deviation between the reconstructed energy E_{rec} and input energy E_{true}, at 1 PeV. In the right panel, the energy response function for showers at impact parameters from 120 to 130 m from the telescope. The energy estimator is the total charge measured in the shower image in unit of number photo-electrons (N_{pe}).

2.3 New Physics Search: LIV Tests

If Lorentz invariance was violated (LIV), photon is no longer a stable particle and could decay on the journey from the source to LHAASO at the Earth. Two possible channels of decay are into either electron/positron pair or 3 γs. The former does not depend how far the source is and the first order violation had been ruled out by previous observations. The second order effect still needs to be tested. The later channel has stronger effect at the second order effect (there is no contribution to the first order expansion) and depends on the length of the path that the original γ-ray traveled. Taking the fact that the SED of the Cygnus region does not show any cut off at the highest energy detected at 1.4 PeV, the violation would not happen to cause the decay of photons below a conservatively selected E_{cut} = 1.14 PeV over the distance of the possible source at $L \sim 1.4$ kpc from the Earth. This results in an lower limit for the second order LIV effect of $3.33{\times}10^{19} L^{0.1} E_{cut}^{1.9} = 2.25 \times 10^{25}$ eV. This is the highest lower limit for the second order LIV effect[7] and still about 3 order of magnitudes below the Planck scale. The limits set by recent experiments and different methods are summarised in Figure 4.

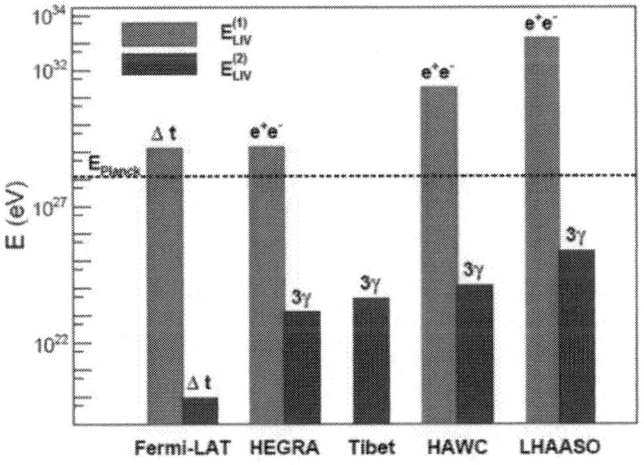

Figure 4. Limits, above which LIV could happen, set by various experiments by measuring the time lag between photons at different energies (LAT) and the photon decay through electron/positron pair or 3 γs (others). The first order expansion of LIV effect has been found a limit much higher limit than Planck scale, indicated by the dotted line. LHAASO which measures the SED of a source in Cygnus region up to 1.4 PeV, pushes the limit to > 10^{33} eV. Giving the distance ~1.4 kpc of the source, the limit for the second order LIV effect has been raised by more than 1 order of magnitude than previous experiments, see [7] and references in it for more detailed discussion.

3 Summary and Prospects

LHAASO as a dual-purpose large scale γ-ray/CR observatory has been built in July 2021 and operated with a very stable high duty cycle since then. By measuring photons around 1 PeV from the Crab, the highest energy photon at 1.4 PeV from the Cygnus region and discovering 12 PeVatrons in the Milky Way, LHAASO has started the era of UHE γ-ray astronomy. As the sole identified origin of the UHE photons, the Crab is accelerating electron/positron pairs to energies at least 2.3 PeV, thus posing challenge to the particle acceleration theories. The SED that extends to 1.4 PeV of a remote source at ~1.4 kpc offers opportunity to test on LIV and set highest limits on the effect.

Multiple EAS parameters are measured with unprecedented precision and statistics in LHAASO, thus opens an opportunity to measure the knees of spectra of individual CR species, such as the protons and iron nuclei at first time. In the first phase of operation of the hybrid observation, 70 million events have been collected for the proton spectrum measurement. The analysis is undergoing. The second phase operation with the full scale LHAASO array started in the dry season of 2021. To accumulate sufficient statistics for the iron spectrum measurement above 10 PeV, a stable operation for 5 or 6 years is necessary.

The CR propagation through the Milky Way will be investigated through the diffuse γ-ray measurement by LHAASO. Some UHE γ-ray sources are investigated in depth to find their CR origin and the potential accelerator, such as the Cygnus region. A much larger catalog of VHE/UHE sources found by LHAASO will be published soon. Among the sources, many CR origin related investigations will be carried out afterwards.

Acknowledgments The author would like to thank all staff members who work at the LHAASO site above 4400 meter above the sea level year round to maintain the detector and keep the water recycling system, electricity power supply and other components of the experiment operating smoothly. We are grateful to Chengdu Management Committee of Tianfu New Area for the constant financial supports to the researches with LHAASO data. This research work is also supported by the National Key R&D program of China under the grant 2018YFA0404201.

References

[1] Z. Cao, Nature Astronomy **5**, 849-849 (2021)
[2] LHAASO Collaboration, Science **373**, 425 (2021)
[3] LHAASO Collaboration, Nature, **594**, 33-36 (2021)
[4] LHAASO Collaboration, ApJL,**917**, L4 (2021)
[5] LHAASO Collaboration, ApJL,**919**, L22 (2021)
[6] LHAASO Collaboration, PRL, **126**, 241103 (2021)
[7] LHAASO Collaboration, PRL, **128**, 051102 (2022)

KM3NeT/ARCA: status of construction and recent physics results

Simone Biagi[1,*] on behalf of the KM3NeT Collaboration

[1]INFN – Laboratori Nazionali del Sud, Via S. Sofia 62, 95123 Catania, Italy.

Abstract. KM3NeT is a multi-site detector devoted to the detection and study of cosmic neutrinos and their sources in the Universe, and to the measurement of the neutrino oscillation parameters. Two underwater detectors are under construction in the Mediterranean Sea, ARCA (Portopalo di Capo Passero, Italy) and ORCA (Toulon, France), optimised respectively for neutrinos in the energy range of 1 TeV–100 PeV and 10 GeV–10 TeV. The mass construction of the detectors has started, and a long-term plan for the completion is in place. At the time of writing this article, 21 (14) detection units are already in operation in the ARCA (ORCA) site from a total of 230 (115). In these proceedings, the major milestones achieved for the construction of the ARCA telescope are discussed. The main physics results obtained with ARCA, in its partial configurations are reported. Finally, an overview of the expected performances of the full detectors will be given.

1 Introduction

Neutrino astronomy is nowadays a well-defined field in astroparticle physics. There are strong expectations around the neutrino experiments which are optimised to detect particles with high energies, $E_\nu > 1$ TeV. Given their neutral electrical charge, neutrinos are not deflected by interstellar magnetic fields but point back to their sources, allowing to identify and study the emitting astrophysical objects using neutrinos as a probe. The cosmic ray energy spectrum has been measured up to energies of 10^{20} eV, however the nature and position in the Universe of the cosmic ray accelerators are still unknown. Recently [1], the IceCube Collaboration published an improved analysis of 10 years of data, reporting the evidence for neutrino emission from the barred spiral galaxy NGC 1068, also known as Messier 77. This is the second neutrino source reported by IceCube after the blazar TXS 0506+056, which was previously identified using a multimessenger approach [2]. The detection of cosmic neutrinos has been interpreted as the "smoking gun" that hadronic processes contribute directly to the cosmic ray acceleration.

Multimessenger astronomy has started to play an emerging role in astroparticle studies. The possibility of joint detections using different mediators, like photons, gravitational waves, high energy cosmic rays, and neutrinos, has already opened new perspectives in cosmic and astronomical studies [3].

In this exciting context, it is urgent to enlarge the network of neutrino detectors to investigate more deeply the reported evidence for cosmic neutrinos. The KM3NeT Collaboration

*e-mail: biagi@infn.it

wants to play an important role in this endeavour and is currently focusing its activities on the construction of two big underwater neutrino detectors, in the Mediterranean Sea. The ARCA telescope (Astroparticle Research with Cosmics in the Abyss) is designed to detect cosmic neutrinos in the energy range 1 TeV−10 PeV. ARCA is currently under construction at 3500 m depth, 80 km South-East the Sicilian coasts of Portopalo di Capo Passero, Italy. The ORCA detector (Oscillation Research with Cosmics in the Abyss) is optimised to determine the neutrino oscillation parameters, through the detection of atmospheric neutrinos in the energy range 10 GeV−10 TeV. The ORCA site is located 40 km from the French coasts near Toulon, at a depth of 2500 m. These proceedings are focused on ARCA; more details about ORCA can be found in [4–6].

There are three main channels that neutrino telescopes exploit for event selection. The first one is the search for a diffuse cosmic neutrino flux. Neutrino events are collected without any assumption on their directions; an excess of events above the atmospheric neutrino flux is searched for energies higher than $\sim$ 50 TeV [7]. The second channel is the search for point-like neutrino sources in the sky. In this case, a statistical accumulation of events around well-defined positions is expected in the sky map [8]. Finally, a multi-messenger approach can be applied to neutrino studies searching for space-time coincidences with the signals reported by other telescopes even using different cosmic probes.

2 The ARCA neutrino telescope

The detection principle of the KM3NeT detectors exploits the Cherenkov light induced by secondary charged particles generated by neutrino interactions inside or in the vicinity of the telescope. A three-dimensional array of photomultiplier tubes (PMTs), which are the sensitive elements of the detector, are distributed in a large volume of a transparent medium like seawater. Deep seawater has the advantage of small light scattering, which improves the angular resolution of the telescope. The main sources of optical background are the decay of ^{40}K in marine water and the bioluminescence, but considering the different energies that characterise these events with respect to the cosmic neutrinos, this background can be easily taken under control. Being located in the Northern Hemisphere, ARCA provides a complementary view to IceCube (located at the South Pole). In addition, the presence of the Galactic Centre in its field of view will open an observative window on a possible population of Galactic sources.

2.1 The seafloor network

The layout of ARCA is shown in Fig. 1 left. Two main electro-optical cables (MEOC in the picture) provide connectivity and power to the offshore infrastructure. These two MEOCs are 100 km long and are connected to the onshore infrastructure, which allocates the power-feeding equipment and the data centre for detector control and data acquisition. The first MEOC embeds a set of 20 optical fibres to establish a data connection between the seafloor network and the shore station. At the submerged end of this MEOC, a primary node (CTF-1) is installed, which hosts a 10 kW medium voltage converter, and 5 electro-optical output ports. The primary node feeds the so-called Phase-1, see Fig. 1 right, via three secondary nodes called Junction Boxes (JB).

The second MEOC embeds a set of 48 optical fibres; two of these fibres are control lines dedicated to the power system. The cable end is split into two secondary branches through a Y-node. The first branch allocates 22 optical fibres and is terminated with a primary node (CTF-2) designed to power the so-called Phase-2. The CTF-2 was installed on the seafloor

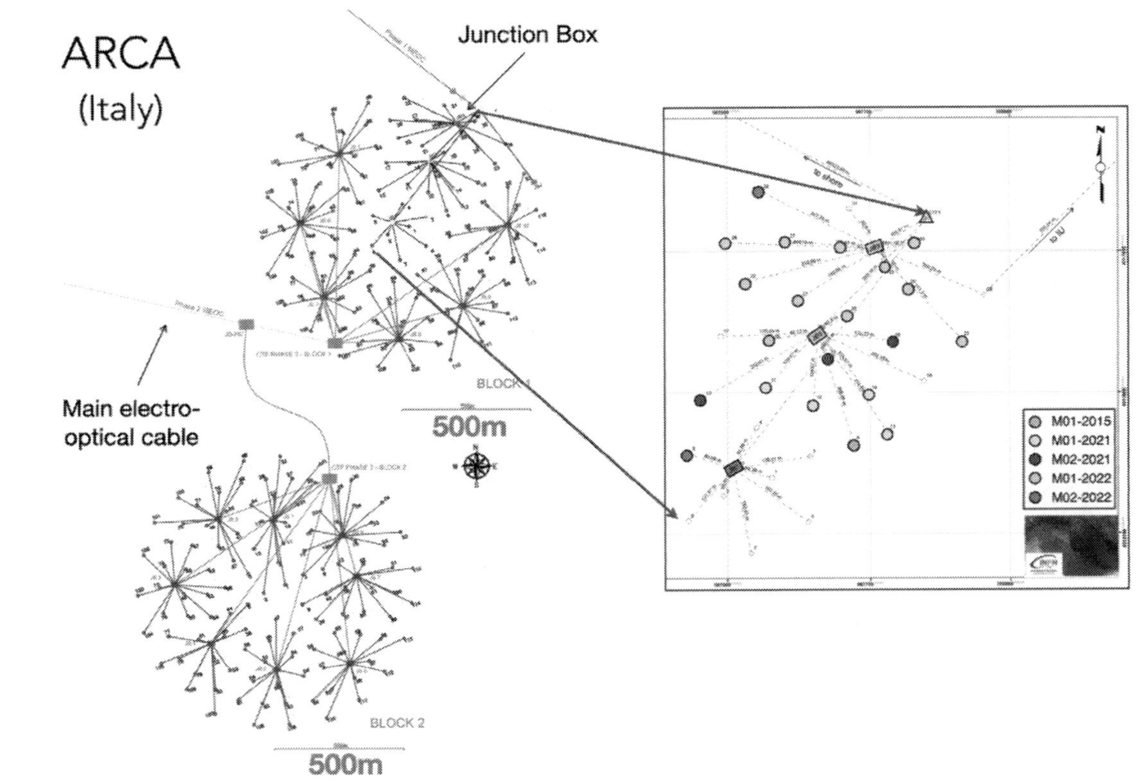

Figure 1. Left: Layout of the completed ARCA detector. Two building blocks of 115 detection units are connected to shore via a network of submarine electro-optical cables and Junction Boxes. Two main electro-optical cables are installed on the seafloor and connected to the onshore laboratory, for power feeding and data acquisition. Right: A zoom of the so-called ARCA Phase-1. The installed detection units (Junction Boxes) are indicated with a coloured circle (rectangle). The colour scheme follows the sequence of marine campaigns indicated by a progressive number and the year.

on November 2022. The remaining 44 fibres are embedded in the second branch, which currently is not equipped at its end with any active termination. The procedure for purchase, construction and installation of the CTF-3 is starting at the time of writing this paper. By design, the CTF-2 will be connected with six JBs, while the CTF-3 will allow for connecting eight JBs. The three CTFs will allow for power and control of the full detector.

2.2 The Junction Box

The ARCA Junction Box has been designed and built to distribute electrical power and an optical fibre channel to the active elements of the detector, which host the PMTs. Its technology was developed by KM3NeT members in synergy with external companies from oil&gas. The goal of the entire JB project was focused on the top-level quality to operate at 3500 m depth during an estimated lifetime of 20 years without any maintenance [9]. The block diagram of the JB is presented in Fig. 2 left. The main core of the JB is the Optical and Power Assembly (OPA), an ensemble of electronics and optical components dedicated to:

- provide power to the various elements of the JB itself;
- provide power to the output ports for the elements of the ARCA telescope;

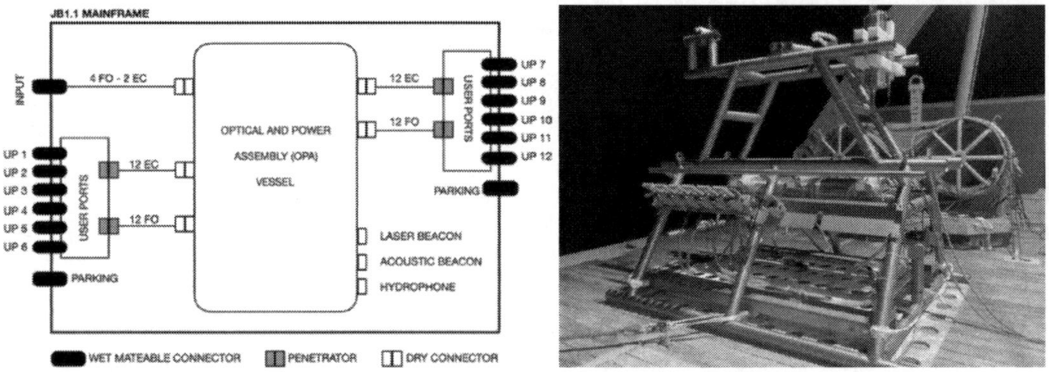

Figure 2. Left: The block diagram of the ARCA Junction Box. A titanium mainframe hosts all the instrumentation and connectors. To improve reliability, the number of connections in water is minimised. The sub-systems, the control and power electronics plus the optical circuits are fully contained in a unique cylinder with a volume of about 100 litres. Right: A JB ready for deployment during a sea campaign. The ROV mateable connectors (in orange) are visible in the front panel; they provide input and output connectivity underwater.

- control and amplification of the optical signal received/sent from/to shore;
- receive the data flow from the telescope and transmit it to shore;
- operate the various instruments mounted on the JB.

The electro-optical system has been designed with all functions intrinsically redundant, where components are duplicated in case of breaking. The electronics boards have been produced with high-reliability components like space, defence or automotive grade.

The electronics is allocated inside a pressure vessel made of titanium grade 2, typically used in this deep-underwater applications. The connectivity with the external world is provided through titanium components produced with high-reliability standards. Dedicated ROV (Remotely Operated Vehicle) mateable connectors[1] are mounted on the mechanical structure of the JB and used as external user ports (Fig. 2 right). The connection to the detection units of the telescope is provided via interlink cables, oil-filled to compensate for the external pressure.

2.3 The detection unit

The detection unit (DU) is the active part of the neutrino detector. It is a vertical structure anchored on the seabed and kept vertical by the buoyancy of its elements. In the ARCA configuration, the DU is 700 m height and hosts 18 Digital Optical Modules (DOMs) [10] with an interspace of 36 m between them. The DU base is a mechanical structure that lies on the seafloor and is equipped with the electronics needed to power and control the DOMs. Two thin Dyneema® ropes are firmly connected to the base and are used to keep the 18 DOMs in place. A thin backbone tube equipped with optical fibres and copper wires for data and power transmission, respectively, runs along the full length of the mooring.

To allow safe transportation and deployment, the DU is folded around a dedicated spherical vehicle at the end of its integration process [11]. Later, the DU is deployed from a boat on the seabed using this spherical structure, as can be seen in Fig. 3. With a precision of about

[1]For this application, we use the Teledyne ODI Nautilus Rolling seal Hybrid (NRH) connectors, see http://www.teledynemarine.com/odi/.

Figure 3. A detection unit enters into the sea. It will go down 3500 m and be deployed on the seafloor. Using a ROV it will be connected to a Junction Box through an electro-optical cable. Finally, the unfurling of the spherical structure will be triggered and the DU will get its final vertical position.

1 m, the DU is placed on the seafloor using an acoustic positioning system. With the support of an ROV, the DU is connected through the interlink to the seafloor network. Once having passed a complete system check performed from shore, the ROV triggers the unfurling mechanism of the vehicle. At the end of this process, the DU gets its final vertical position and the launcher vehicle is finally recovered on the sea surface to be reused.

The sensitive element of the telescope is the Digital Optical Module, a 0.44 m diameter pressure-resistant glass sphere which contains 31 3 inches photomultiplier tubes [12]. The DOM also hosts calibration devices and electronic boards for power, readout and data acquisition [13]. The implemented technology allows for sub-nanosecond timing accuracy of the PMT signal. The dynamic range spans from the single photon up to a cascade of thousands of photons, providing additional information useful for the reconstruction of the energy of the event. Given the presence of 31 PMTs, the DOM has an angular acceptance close to 4π, and can provide a better measurement of the photon direction for neutrino track reconstruction.

The current configuration of ARCA is shown in Fig. 1 right: 21 detection units are in place and taking data. Typically, two sea campaigns per year are organised. The mass production of the detector started in 2021, once the main technical issues were addressed and solved. In Fig. 4, the timeline of the construction of ARCA is reported. The first DU was installed in December 2015; it is still in operation and taking data. Data taking was put on hold during two periods, the former related to an electrical problem with the seabed infrastructure and the latter due to a major upgrade of the shore station. The issues with the infrastructure were solved with a new design of the Junction Box, see Sect. 2.2.

3 Some selected results

3.1 Neutrino point source search

The first analysis [14] devoted to the search for point-like neutrino sources is performed on a data sample equivalent to 92 days of operation, taken with ARCA in a six-line configura-

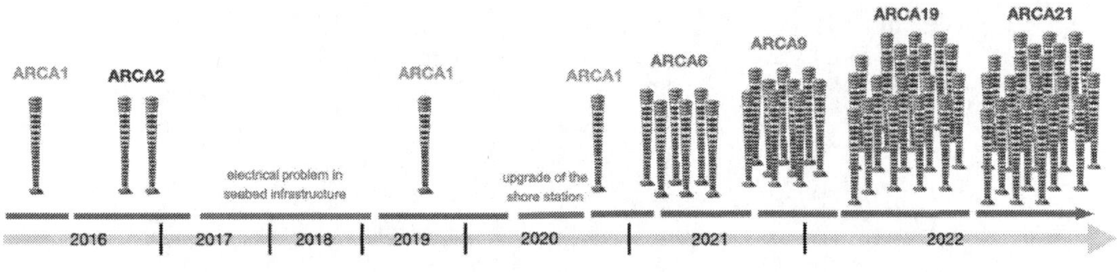

Figure 4. The ARCA detector is growing following the integration activities of the various elements, that are deployed during dedicated marine campaigns. Typically, two campaigns per year are performed, avoiding the winter period. Even if in partial configuration, ARCA is continuously taking data. Two long shutdown periods are reported in red: the first due to an electrical issue in the seabed infrastructure that was fixed with a complete redesign of the Junction Boxes, and the second related to a major upgrade of the shore station which was enlarged to support the full detector with two building blocks.

tion. The neutrino emission is searched among 40 candidate point sources plus 6 extended sources; the 46 candidates are reported on a sky map in Fig. 5 left. A binned likelihood search is applied to the data set corresponding to the period between May and September 2021. Background-only and signal+background samples are built from scrambled real data and Monte Carlo simulations. The neutrino signal is generated with an E^{-2} energy spectrum in a 5° cone around the source.

The most significant source corresponding to the lowest p-value (0.02) is the radio galaxy Centaurus A (RA= 201.36, $\delta = -43.01$), for which 2.6 signal events are fitted. The source is indicated in Fig. 5 left with a yellow arrow. The p-value of $\sim$ 2% is compatible with the expectation from the pure background hypothesis. Upper limits on the neutrino fluxes for the 46 candidates are evaluated and reported as blue points in Fig. 5 right, together with the ARCA 6-line sensitivity. As a comparison, the ANTARES and IceCube sensitivities are also reported, together with the full ARCA detector expected sensitivity.

3.2 Full detector performances

The complete ARCA detector will be composed of 230 detection units, distributed in two building blocks of 115 DUs each. Dedicated algorithms have been developed in order to reconstruct the neutrino direction and energy. The information of the time and charge of the signal collected by the PMTs is used as an input of these algorithms, together with the real-time position of the PMTs triangulated with an acoustic positioning system.

The neutrino events can be classified into two main topologies: track-like and shower-like events. Track-like events are generated by charge current interactions of ν_μ and are the ideal tool for astronomy, as the secondary muon can travel several kilometres in rock and seawater producing a clean Cherenkov light cone. The angular resolution of ARCA is expected to be better than 0.2° at energies above 10 TeV (Fig. 6 left). For the same events the energy resolution, evaluated as the FWHM of the $\log_{10}(E_{reco}/E_\mu)$ distribution, will be $\sim$ 0.25.

The shower-like events are produced by neutral current interactions of ν of any flavour, plus charge current interactions of ν_e and ν_τ. In this case, events are typically contained within the detector active volume, hence the energy resolution is improved with respect to track-like events being at the level of 5%. On the other hand, the angular resolution in the same energy region is expected to be $\sim$ 2°, which is worse than for track-like events.

Another important physics goal of KM3NeT is the detection of supernova neutrinos. The KM3NeT telescope sensitivity to Galactic supernovae and its associated online alert system

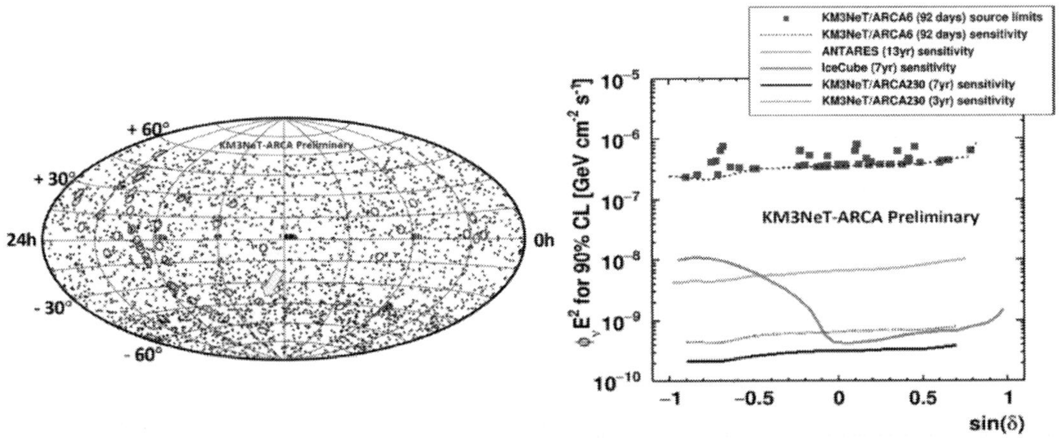

Figure 5. Left: The neutrino sky map of the ARCA 6-line detector with candidate sources indicated with red circles. The yellow arrow points to the source with the best p-value, see text. Right: Upper limits on the cosmic neutrino flux for the candidate sources are reported as blue points. The ARCA sensitivity is displayed as a dashed line and compared to ANTARES and IceCube sensitivities.

for multi-messenger studies was presented during the conference, see [15]. From the time profile of the associated neutrino emission, it will be possible to derive the supernova time-evolution and improve the models that explain how these objects work.

Also at this conference, the analysis of the search for starburst galaxies using the full ARCA detector was presented, see [16]. Using a point-like search approach, the most promising local starburst galaxies are selected and the expected sensitivity of the full ARCA detector to observe point-like neutrino excesses is evaluated. ARCA will have the power to constrain several phenomenological scenarios, providing evidence of the link between star-forming processes and hadronic emissions.

4 The next future

The KM3NeT Collaboration has started the mass construction of ARCA. The integration capability of the various elements of the detector has improved, and a distributed production model has been implemented for the delivery of the full detector, which will be composed of two building blocks in 2030. The KM3NeT Collaboration has currently collected the funds required to complete the first building block and start the construction of the second one. These funds will allow to build ~130 detection units, the related seafloor infrastructure, and the onshore lab facilities. In the meantime, data will be analysed with the detector in partial configurations, waiting for the final completion.

References

[1] R. Abbasi et al. (IceCube Collaboration), *Evidence for neutrino emission from the nearby active galaxy NGC 1068*, Science **378**, 538 (2022).

[2] M.G. Aartsen et al. (IceCube Collaboration), *Evidence for high-energy extraterrestrial neutrinos at the IceCube detector*, Science **342**, 947 (2013).

[3] A. Albert et al., *Search for multimessenger sources of gravitational waves and high-energy neutrinos with Advanced LIGO during its first observing run, ANTARES, and IceCube*, ApJ **870**, 134 (2019).

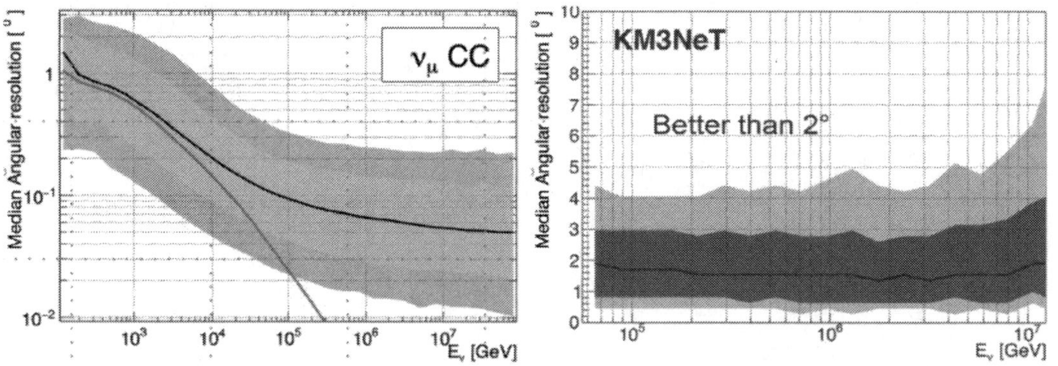

Figure 6. Left: Median angular resolution of the full ARCA detector as a function of the neutrino energy for track-like events. Shadowed areas represent the 50 and 90 percentile. The red line is the angle between the ν_μ and the secondary μ. Right: Median angular resolution of the full ARCA detector as a function of the neutrino energy for shower-like events.

[4] S. Adrián-Martínez et al. (KM3NeT Collaboration), *Letter of intent for KM3NeT 2.0*, J. Phys. G: Nucl. Part. Phys. **43**, 084001 (2016).

[5] S. Aiello et al. (KM3NeT Collaboration), *Determining the neutrino mass ordering and oscillation parameters with KM3NeT/ORCA*, Eur. Phys. J. C **82**, 26 (2022).

[6] S. Aiello et al. (KM3NeT Collaboration), *Sensitivity to light sterile neutrino mixing parameters with KM3NeT/ORCA*, JHEP **10**, 180 (2021).

[7] A. Albert et al. (ANTARES Collaboration), *All-flavor search for a diffuse flux of cosmic neutrinos with 9 years of ANTARES data*, ApJL **853**, L7 (2018).

[8] S. Aiello et al. (KM3NeT Collaboration), *Sensitivity of the KM3NeT/ARCA neutrino telescope to point-like neutrino sources*, Astropart. Phys. **111**, 100 (2019).

[9] F. Ameli et al., *Underwater node for electrical power distribution and optical data transmission for multidisciplinary science in the Mediterranean Sea*, 2022 IEEE International Workshop on Metrology for the Sea, 441–445 (2022), DOI: 10.1109/MetroSea55331.2022.9950925.

[10] S. Aiello et al. (KM3NeT Collaboration), *The KM3NeT multi-PMT optical module*, JINST **17**, P07038 (2022).

[11] S. Aiello et al. (KM3NeT Collaboration), *Deep-sea deployment of the KM3NeT neutrino telescope detection units by self-unrolling*, JINST **15**, P11027 (2020).

[12] I.C. Rea (for the KM3NeT Collaboration), *The multi-PMT optical module of KM3NeT*, these proceedings.

[13] S. Biagi et al. (for the KM3NeT Collaboration), *The data acquisition system of the KM3NeT detector*, PoS(ICRC2015)**1172** (2015).
T. Chiarusi et al. (for the KM3NeT Collaboration), *The KM3NeT data acquisition system – Status and evolution*, these proceedings.

[14] R. Muller et al. (for the KM3NeT Collaboration), *Search for cosmic neutrino point sources and extended sources with 6 lines of KM3NeT/ARCA*, DOI: 10.5281/zenodo.6805394 (2022).

[15] M. Bendahman (for the KM3NeT Collaboration), *Neutrinos from core-collapse supernovae at KM3NeT*, these proceedings.

[16] W.I. Ibnsalih (for the KM3NeT Collaboration), *KM3NeT/ARCA Sensitivity to Starburst Galaxies*, these proceedings.

TeV emission from Gamma Ray Bursts, checking the hadronic model

Dafne Guetta[1,*], *Silvia* Gagliardini[2,3], *Silvia* Celli[2,3], *Angela* Zegarelli[2,3], *Antonio* Capone[1,2], *Stefano* Campion[2,3], and *Irene* DiPalma[2,3]

[1]Department of Physics, Ariel University, Ariel, Israel
[2]Dipartimento di Fisica dell'Università La Sapienza, P. le Aldo Moro 2, I-00185 Rome, Italy,
[3]Istituto Nazionale di Fisica Nucleare, Sezione di Roma, P. le Aldo Moro 2, I-00185 Rome, Italy,

Abstract. Gamma-Ray Bursts (GRBs) are the most luminous explosions in the Universe. Their luminous prompt emission makes them detectable from cosmological distances. Most GRBs have been detected below a few MeV, however at least a hundred GRBs have been detected at high (0.1 GeV) energies and observed up to tens of GeV with the *Fermi* Large Area Telescope (LAT). Some GRBs have been observed at (0.1–1) TeV by ground-based imaging atmospheric Cherenkov telescopes. To date, the high energy emission mechanism is not understood. In this paper we review the possible leptonic and hadronic mechanisms capable of producing the ~ TeV emission detected in GRBs. In paricular we concentrate on the hadronic origin of this radiation component and discuss in detail the numerical simulation elaborated to reproduce the observed sub-TeV observations of GRB 190114C.

1 Introduction

The detection of GRBs in sub-TeV gamma rays has disclosed a new window of the electromagnetic spectrum for studying mildly relativistic transient outflows at the highest energies. GRB 190114C is the first long-duration GRB reported in the TeV band by the Major Atmospheric Gamma Imaging Cherenkov (MAGIC) telescopes. High-energy gamma rays (0.2–1 TeV) were detected by the instrument with high significance [1], during both the prompt and afterglow phases of the GRB emission. The sub-TeV radiation component has a power comparable to that of the sub-GeV one, attributed to the synchrotron process. Therefore, sub-TeV emission has been interpreted as due to leptonic processes [2], usually considered responsible for gamma ray production in GRBs. However, a purely leptonic scenario consisting of synchrotron self-Compton radiation produced at external shocks has been shown to not match the broadband emission observed in GRB 190114C [2]. In addition, the large number of parameters involved in the model highly limits its prediction power. Although firm conclusions on the production mechanisms of GeV-TeV emission have been reached so far [3–9], the hypothesis that part of this emission might be caused by the presence of a hadronic component cannot be excluded. Several authors [4, 6, 7] have hence explored the possibility that processes involving photo-meson interactions might play a relevant role in the formation of the gamma-ray spectrum of GRB 190114C up to TeV energies. Since the

*e-mail: μdafneguetta@gmail.com

dissipation mechanism responsible for the acceleration of electrons could be invoked for the acceleration of protons, high-energy neutrinos are also expected [10, 12–14]. In this review we provide a concise summary of the radiation mechanisms that may explain the spectral and temporal properties of the $\sim TeV$ emission in GRBs. In particular we test a scenario in which the very high energy (VHE) emission of GRB 190114C is due to photo-hadronic interactions occurring in the GRB jet during the so-called prompt phase. To this extent, we consider the standard fireball model in which, within the relativistic outflow of the jet, energy dissipation occurs at shocks internally to the jet or through interactions with ambient matter [10, 12]. Thus, a substantial part of the bulk kinetic energy is converted into internal energy, which is then distributed between electrons, protons, and the magnetic field. The internally accelerated electrons are presumably responsible for the keV–GeV photons observed in the GRB, which are emitted through synchrotron or inverse Compton processes. Accelerated protons may interact with these photons ($\sim$ MeV) and produce both neutral and charged pions, which in turn decay, originating respectively high-energy photons and neutrinos. The review is structured as follows: In Sec. 2 we discuss several high energy radiation mechanisms. In Sec. 3 we introduce the spectral and temporal properties of GRB 190114C. In Sec. 4 we present our hadronic model. In Sec. 5 we compare the photon flux resulting from our simulation to the Extragalactic Background Light (EBL) deconvolved MAGIC data, obtaining a set of best fit values of the model parameters. From the same simulation we predict a flux of high energy neutrinos and we evaluate the expected number of events in present and future experiments. A discussion of the results is provided in Sec. 6, where conclusions are ultimately derived.

2 Mechanism models for the high energy emission

In this section we consider different high energy mechanism that can explain the $\sim$ TeV emission observed. This emission may be due to a completely leptonic process like the synchrotron emission by electrons accelerated in the shock region. However several authors (i.e. [6, 11]) have shown that the maximum energy that can be reached with this mechanism is $\sim$ 10 GeV. Another leptonic model is the synchrotron self-compton emission by the same electron population that produced the seed synchrotron radiation. A distribution of relativistic electrons can inverse-Compton scatter some of the same synchrotron photons that it produced, leading to a synchrotron self-Compton emission. With this model it is hard to explain both the low and the high energy GRB emission. Another mechanism may be the External inverse-Compton emission that arises when the seed photons are inverse Compton scattered to high energies by relativistic electrons in a location different than where the seed photons were produced. This seems to be the best leptonic model to fit the data to date [6]. Alternatively, hadronic models may reproduce the high high energy emission. One of these models is the synchrotron emission produced by high-energy protons that are accelerated at shocks (like the electrons). However, the emitted power per particle is much smaller with respect to that for electrons and therefore this process seems to be unlikely. Another possible model is the photo-hadronic model where the high energy photons come fro neutral pion decay. We discuss this model in detail in the following sections.

3 GRB 190114C: spectral and temporal properties

GRB 190114C was the first GRB detected at sub-TeV energies, among the few observed so far. The combination of a rapid follow-up by MAGIC [15] with the rather contained distance of the source (confirmed from its optical counterpart to be at redshift $z = 0.4245$ [16, 17]), in terms of extra-galactic scales, has allowed to unveil the presence of an extremely energetic radiation component in GRBs, long before theorised [18]. A wealth of observations has allowed for the precise characterisation of the temporal and spectral features of this source. In the gamma-ray band, starting from the triggering instrument Swift [19], further observations have followed by GBM [20] and LAT [21] onboard of the *Fermi* satellite (up to 22.9 GeV), AGILE/MCAL [22], Integral/SPI-ACS [23] and Konus-Wind [24].

The prompt phase of GRB 190114C, as observed by Fermi-GBM, appears as a multi-peak emission lasting $T_{90} \simeq 116$ s (50-300 keV), being T_{90} the duration where 90% of the gamma-ray fluence was detected. The time-averaged spectrum in the first $\sim$ 40 s can be described by a Band function with low and high-energy slopes equal to $\alpha = 1.058$ and $\beta = 3.18$, respectively, ans a break energy value of $E_b \simeq 1.1$ MeV in the observed frame [20], and a gamma-ray fluence of $F_\gamma = 3.99 \times 10^{-4}$ erg cm^{-2} (10-1000 keV). As such, the isotropic energy release in the source frame amounts to $E_{\rm iso} \simeq 3\times 10^{53}$ erg, indicating a fairly energetic (though not the most extreme) GRB. In turn, MAGIC reported a first detection starting from 68 s after the Fermi-GBM trigger with a significance above 50σ: VHE emission lasted for $\sim$ 40 minutes, with the highest photon energy observed amounting to $E_{\rm max} = 0.852$ TeV [1]. Within this extended temporal window, time-dependent analysis of VHE data was also performed, showing a systematical decrease in flux normalization as well as a steepening trend over time. The earliest high-energy photon spectrum reported by MAGIC refers to the time interval 68-110 s, that entirely overlaps with the T_{90} estimated by Fermi-GBM for the prompt emission. During this time interval, the intrinsic burst spectrum in the 0.2-1 TeV band is characterised by a pure power-law ($\propto E^{-\xi}$) with $\xi = 2.16^{+0.29}_{-0.31}$ [2]. The photon spectral slope at the source has been derived taking into account the severe attenuation of the gamma-ray flux due to its propagation to the Earth within the EBL [2], according to the Dominguez et al. model [25]. We hence consider such an intrinsic spectrum and we compare it with the prediction of the gamma-ray flux emerging from our simulation of the IS region. Additionally, we have investigated the effects of adopting a different EBL model, e.g. that from Franceschini et al. [26, 27], finding results consistent with the Dominguez et al. model within the statistical uncertainty of the MAGIC measurements.

The temporal overlap between the MAGIC observations and T_{90} is not sufficient to definitely attribute the high-energy photons measured by MAGIC to the prompt or to the afterglow phase of the GRB emission. In this work, we consider a scenario where the observed TeV emission can be ascribed to the prompt phase, in particular to radiation emerging from hadronic interactions occurring at the IS region between accelerated protons and the Band-like target radiation field. The two main features of a hadronic scenario are the bulk Lorentz factor Γ, and the amount of energy channeled into relativistic protons $E_{\rm iso,p}$. The latter can be expressed with the baryon loading f_p defined as $f_{\rm p} = E_{\rm iso,p}/E_{\rm iso}$. Both the values of Γ and $f_{\rm p}$ can be constrained in order to reproduce the VHE spectrum in terms of hadronic gamma rays. Concerning the value of Γ, we estimate the minimum Lorentz factor required by a $E_\gamma \simeq 1$ MeV photon to be revealed at Earth, as observed by Fermi-GBM. This can be constrained by requiring that the optical thickness τ for pair production $\gamma\gamma \longrightarrow e^+e^-$ has to be $\tau \leq 1$. Following Eq. (5) of [28], we computed the total number of photons emitted during the first $\sim$ 40, s, with energy greater than $E_{\rm max,pair} = (2\Gamma m_e c^2)^2/E_\gamma$, for different variability times $t_{\rm var} = 6, 3, 1$ ms [29], obtaining minimum Lorentz factor values $\Gamma_{\rm min} \simeq 60, 70, 90$ respectively.

4 The hadronic model: numerical simulations

We consider a model where high-energy protons, accelerated in the shock region, interact with photons, producing charged and neutral pions [13]. The pion decay products include leptons and photons:

$$\pi^+ \rightarrow \mu^+ + \nu_\mu \rightarrow e^+ + \nu_e + \bar{\nu}_\mu + \nu_\mu \,, \tag{1}$$

$$\pi^- \rightarrow \mu^- + \bar{\nu}_\mu \rightarrow e^- + \bar{\nu}_e + \nu_\mu + \bar{\nu}_\mu \,, \tag{2}$$

$$\pi^0 \rightarrow \gamma + \gamma \,. \tag{3}$$

Modelling the physical processes occurring inside the IS region of the expanding GRB fireball requires the characterisation of the site where particles propagation and interactions take place. The emitted flux depends strongly on the GRB lorentz factor, Γ and to the variability time t_{var}. In our simulation, we assume for the variability timescale values ranging from 1 to 6 ms, the latter being suggested by observations performed during the prompt phase of GRB 190114C [29]. In turn, the Γ factor is treated as a free parameter of the model, whose most likely value will be fixed as the one that best reproduces MAGIC data.
All the details of the MC calculation are reported in [30]. The kinematics of $p\gamma$ interactions follows the methods outlined in [31]. The energy particle distribution for interacting protons, charged pions, muons, and neutrinos is shown in Fig.1 of [30].

Concerning the neutral pion production and decay, we consider the possibility that each originated gamma ray might further interact within the IS shell with low energy ambient photons, via $e^{\pm}$ pair production. For the purposes of our simulation, the Band spectrum of target photons is assumed to extend outside of the Fermi-GBM observed energy range, by about three decades, as to allow the pair production to occur with the gamma ray originated from π^0-decay. As a result of our computation, we find the fraction of absorbed high energy photons amounts to $\simeq 60\%$ for 10 GeV gamma rays in the IS frame ($\Gamma = 100$ and $t_{\mathrm{var}} = 1$ ms). This fraction increases for more energetic gamma rays. We observe that, at fixed gamma-ray energy, the effect of $\gamma\gamma$ absorption results larger for lower values of Γ, since the target photon density increases lowering the bulk Lorentz factor.
The gamma rays surviving the $e^{\pm}$ production process in the IS shell will emerge from this region and would eventually be observed in the laboratory frame with energy $E_{\mathrm{obs}} = \Gamma E_{\mathrm{IS}}/(1+z)$. Such photons can be directly compared with the intrinsic Spectral Energy Density (SED) derived by MAGIC.

5 Results: gamma rays and neutrinos from photo-hadronic interactions

We perform MC simulation for different values of t_{var} and Γ, obtaining spectra of high-energy gamma rays and neutrinos emerging from the interaction region. We then convert these spectra into expected fluxes on Earth by taking into account the cosmological nature of GRBs. For the particle energy flux $E^2\phi(E)$, either photons or neutrinos, we rescaled the IS energy spectrum through: i) the comoving distance $d_{\mathrm{c}} = d_{\mathrm{L}}/(1+z)$, where d_{L} represents the luminosity distance, ii) the first time interval of MAGIC observations $\Delta t = 42$ s, and iii) the energy conversion factor from the IS to the observed frame, as:

$$E^2\phi(E) = f_{\mathrm{p}}\Gamma \frac{(1+z)}{4\pi d_{\mathrm{L}}^2 \Delta t}\left(E^2 \frac{dN}{dE}\right)_{\mathrm{IS}} \,. \tag{4}$$

For each simulated Γ and t_{var} set of values, the baryon loading f_{p} has been varied in the interval $0.1 - 5.0$ in uniform steps of width $\Delta f_{\mathrm{p}} = 0.1$, in order to derive the best agreement between the EBL-deconvolved MAGIC data in the 68-110 s time interval [2] and the simulated fluxes, through a χ^2 statistical test. Considering the minimum bulk Lorentz factor values dictated by the opacity argument and derived in Sec. 3, we obtain the following set

 https://doi.org/10.1051/epjconf/202328001005

of best-fit parameters: for the case $t_{var} = 6$ ms, we find $\Gamma = 70$ and $f_p = 0.95$; in turn, for $t_{var} = 3$ ms, we obtain $\Gamma = 80$ and $f_p = 0.93$; finally, for $t_{var} = 1$ ms, we derive $\Gamma = 100$ and $f_p = 1.1$. Remarkably, despite the broad variation in the values of bulk Lorentz factor that characterize the models possibly reproducing sub-TeV data, an almost constant f_p is obtained, suggesting a comparable amount of energy being channeled in the hadronic and leptonic jet components. By fixing Γ to the best-fit value, we evaluated the statistical uncertainty on flux normalisation. This result is shown in Figure 1, where the modellings that best reproduce the EBL-deconvolved MAGIC data are reported on top of it, the shaded area representing the 1σ statistical uncertainty evaluated over f_p.

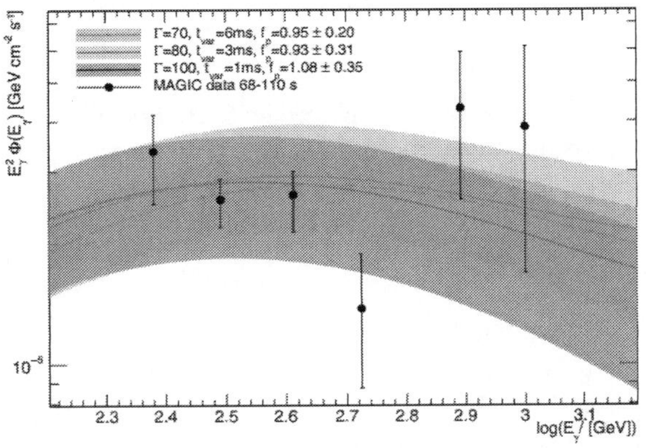

Figure 1. Comparison between the EBL-deconvolved flux of GRB 190114C, as measured by MAGIC in the temporal interval 68-110 s [2], and the simulated photon flux arising from the π^0-decay, after accounting for internal gamma-ray absorption. Different parameter values of the model are tested, as indicated in the legend.

A direct proof of the hadronic origin of the observed TeV radiation might come from coincident neutrino observations. Because of the transient nature of GRBs, the detection of a single neutrino event would allow to identify these sources as extreme hadronic accelerators. Despite the experimental efforts, no clear signal has emerged so far in data from the ANTARES and IceCube telescopes [34] [35], when searching for spatial and temporal coincidences with the prompt emission of GRBs. This leads to a constraint on the possible contribution of standard GRBs to less than 10% of the diffuse cosmic neutrino flux [36–39]. Additionally, it is still unclear whether TeV-emitting GRBs behave as the standard GRB population: consequently, it appears of paramount importance to investigate neutrino emissions from this newly discovered class of sources. Both the ANTARES and IceCube Collaborations have searched for coincident neutrino-induced signals from the direction of GRB 190114C. No events were observed in extended time windows, covering both the prompt and the afterglow phase of the GRB, leading to upper limits on the expected neutrino fluence. In the case of ANTARES, the 90% confidence level integrated limit amounts to 1.6 GeV/cm^2, while for IceCube it amounts to 0.44 GeV/cm^2 [40, 41]. In both cases, the constraints are limited to the muon neutrino component reaching Earth. In fact, because angular precision is a crucial feature to reduce the atmospheric background entering the search cone angle, muon neutrino charged-current interactions constitute the better astronomical channel, as muons emerge from these, which can be identified in large volume neutrino telescopes as long tracks. Neutrino fluxes arriving on Earth differ from those produced at source due to the effect of lep-

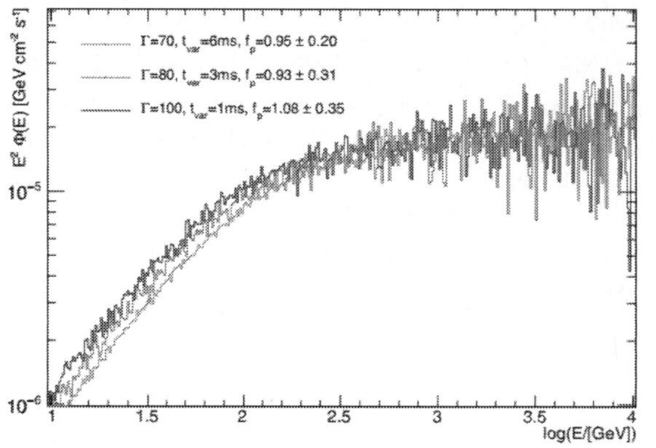

Figure 2. Muon neutrino and antineutrino fluxes expected on Earth from GRB 190114C, obtained from the simulations in the time interval 68-110 s for different model parameter values, as indicated in the legend.

tonic flavor mixing: hence, in order to have a direct comparison between the experimental upper limits and the predictions from the hadronic modelling here developed, neutrino oscillations have to be accounted for. By considering the normal ordering case in the standard three-flavour scenario [42], we obtain the neutrino energy flux on Earth, which is shown for muon neutrinos and antineutrinos in Fig. 2. Current neutrino detectors ANTARES and IceCube have investigated their data looking for possible spatially coincident neutrino signals, however the resulting upper limits are a few order of magnitudes above the model predictions here presented, indicating that the non-detection of current instruments is compatible with the hadronic model expectations for the values of $f_{\rm p}$ and Γ that better reproduce the sub-TeV gamma-ray MAGIC data. To understand whether future neutrino detectors, as the full configuration of KM3NeT, might be able to uncover a level of signal comparable to that expected from GRB 190114C, we first obtain the muon neutrino (and antineutrino) fluence $dN_{\nu_\mu}/(dE_{\nu_\mu}dS)$, and then compute the expected number of track-like events as

$$N_{\rm events}(\delta) = \int A_{\rm eff}^{\nu_\mu}(E_\nu, \delta)\left(\frac{dN_{\nu_\mu}}{dE_{\nu_\mu}dS}\right)_{\rm Earth} dE_{\nu_\mu}\,, \tag{5}$$

where the neutrino telescope effective area $A_{\rm eff}^{\nu_\mu}$ is a function of the neutrino energy and of the source declination δ. For the purpose of the computation, the public effective areas of ANTARES, IceCube and KM3NeT are adopted, averaged over neutrinos and antineutrinos. In the case of ANTARES, we consider the declination band $-45^o < \delta < 0^o$ [34] including the position of GRB 190114C, in the case of IceCube instead the effective area refers to the band $-30^o < \delta < 0^o$ [35], while for KM3NeT we adopt mean values of the high-energy detector ARCA (averaged over the entire sky) [43]. It is also worth to note that, while the ANTARES and IceCube have been released at the analysis level for point source studies, the KM3NeT/ARCA one is at trigger level. We provide in the following the expected number of neutrino-induced events in the three large-volume detectors, as reported in Table 1 relatively to the simulation with $t_{\rm var}$ = 1 ms, Γ = 100 and $f_{\rm p}$ = 1.1. Even though the coincident background rate due to atmospheric neutrinos is by orders of magnitude lower than the signal rate expected from a source like GRB 190114C, a detection from a single GRB appears to be very difficult. This result suggests that stacking several GRBs of the same kind could be the only viable solution for a significant detection.

Detector	Declination band	N_{events}
ANTARES	$-45^o < \delta < 0^o$	1×10^{-3}
IceCube	$-30^o < \delta < 0^o$	2×10^{-2}
KM3NeT/ARCA	Average	1×10^{-1}

Table 1. Signal events expected in different neutrino telescopes, as induced by ν_μ interactions during the [68;100] s time interval of the prompt emission of GRB 190114C, for the model with $t_{\text{var}} = 1$ ms, $\Gamma = 100$ and $f_p = 1.1$. The computations refer to instrument effective areas in the declination band including the source location (ANTARES [34], IceCube [35], and KM3NeT [43]).

6 Discussion and Conclusions

The successful observation of radiation from GRBs extending up to the TeV domain has provided further evidence of the extreme nature of these sources. However, there is still poor understanding of the processes that characterize the observed TeV emission emerging from GRB 190114C (and the few other GRBs detected at TeV energies). The newly discovered radiation component might witness the presence of effective hadronic acceleration in the jet of the burst, possibly already during the prompt phase of the emission. This occurrence would establish the connection between GRBs and Ultra-High-Energy Cosmic Rays (UHECRs), which remains a long-standing paradigm still to be proven. To test the hypothesis of the hadronic origin of TeV radiation, we developed a MC simulation of photo-meson interactions between high-energy protons and target photons distributed according to a Band-like spectrum, as indicated by Fermi-GBM observations. After computing the spectra of secondary particles emerging from these interactions, we additionally simulated the electromagnetic absorption that gamma rays undergo in the IS shell. The spectrum of escaping photons thus obtained has been compared to the intrinsic source spectrum derived by deconvolving MAGIC observations of GRB 190114C in the EBL. The free parameters of the hadronic model have thus been constrained by means of a residual minimisation procedure, indicating that the source channelled a comparable amount of energy into photons and hadrons. Extended studies about the entire sample of observed TeV GRBs are required to understand the physical mechanisms responsible for the TeV emission. In particular, it appears to crucial to have a better characterization of the very-high-energy photon spectrum in the early stages of the GRB emission, which seems to be currently limited by the prompt response of imaging atmospheric Cherenkov telescopes in pointing. Confirmation of the hadronic origin of sub-TeV radiation might, in principle, arise from neutrino observations. In the context of the parameters that better reproduce MAGIC data, however, such a detection from GRB 190114C appears extremely unrealistic, as confirmed by the lack of spatial correlations in data from both the ANTARES and IceCube neutrino telescopes. We may conclude that

1. Both leptonic and hadronic interpretations of the TeV data cannot be excluded.
2. Extended studies about the entire sample of observed TeV GRBs are required to understand the physical mechanisms responsible for the TeV emission.
3. In order to have a complete understanding on the mechanism responsible of the high energy emission, it is crucial to have a better characterization of the very-high-energy photon spectrum in the early stages of the GRB emission. This seems to be currently limited by the prompt response of imaging atmospheric Cherenkov telescopes in pointing but it will be improved in the next future

References

[1] V. A. Acciari et al. (MAGIC Collaboration), Nature 575 (2019) 455
[2] V. A. Acciari et al. (MAGIC Collaboration), Nature 575 (2019) 459
[3] M. E. Ravasio, G. et al. A&A 626 (2019) 12
[4] N. Fraija, R. Barniol Duran, S. Dichiara & P. Beniamini, ApJ 883 (2019b) 162
[5] X. Y. Wang, R. Y. Liu, H. M. Zhang, S. Q. Xi and B. Zhang, ApJ 884 (2019) 117
[6] E. Derishev & T. Piran, ApJ, 880 (2019) 27
[7] S. Sahu, & C. E. López Fortín, ApJ 895 (2020) 41
[8] V. Chand, P. S. Pal, A. Banerjee, V. Sharma, P. H. T. Tam, and X. He, ApJ 903 (2020) 9
[9] J. A. Rueda, R. Ruffini, M. Karlica, R. Moradi & Y. Wang, ApJ 893 (2020) 148
[10] D. Eichler, D. Guetta & M. Pohl, ApJ 722 (2010) 543
[11] R. Gill& J. Granot, 2022, Galaxies, vol. 10, issue 3, p. 74
[12] D. Guetta, M. Spada & E. Waxman, ApJ 559 (2001) 101
[13] D. Guetta, JHEAp, 7, 90 (2015)
[14] L. Yacobi, D. Guetta & E. Behar, ApJ 793 (2014) 48
[15] R. Mirzoyan et al., ATel 12390 (2019) 1
[16] J. Selsing et al. (NOT Collaboration) GCN 23695 (2019)
[17] A. J. Castro-Tirado et al. (GTC Collaboration) GCN 23708 (2019)
[18] E. V. Derishev, V. V. Kocharovsky & V. V. Kocharovskya, ASR 27 (2001) 813-818
[19] J. D. Gropp et al. (Swift Collaboration) GCN 23688 (2019)
[20] R. Hamburg et al. (Fermi-GBM Collaboration), 2019, GCN 23707
[21] D. Kocevski et al. (Fermi-LAT Collaboration), 2019, GCN 23709
[22] A. Ursi et al. (AGILE Collaboration), GCN 23712 (2019)
[23] P. Minaev & A. Pozanenko, GCN 23714 (2019)
[24] D. Frederiks et al. (Konus-Wind Collaboration), GCN 23737 (2019)
[25] A. Dominguez et al., Mon. Not. R. Astron. Soc. 41 (2011) 2556–2578
[26] A. Franceschini & G. Rodighiero, A&A 603 (2017) A34
[27] A. Franceschini & G. Rodighiero, A&A 614 (2018) C1
[28] Y. Lithwick & R. Sari, The Astrophysical Journal 555 (2001) 1
[29] M. Ajello et al., ApJ 890 (2020) 9
[30] Gagliardini S., Celli S., Guetta D., Zegarelli A., Capone A., Campion S. and Di Palma I. 2022 arXiv220901940G
[31] M. Fasano, et al. , JCAP 09 (2021) 044
[32] A. Atoyan & C.D. Dermer, Phys. Rev. Lett. 87 (2001) 221102
[33] S. Vernetto and P. Lipari, Phys. Rev. D 94 (2016) 063009
[34] S. Adrian-Martínez et al. (ANTARES Collaboration), Astrophys. J. 760 (2012) 53
[35] M. G. Aartsen et al. (IceCube Collaboration), Astrophys. J. 796 (2014) 109
[36] S. Adrian-Martínez et al. (ANTARES Collaboration), A&A 559 (2013) A9
[37] A. Albert et al. (ANTARES Collaboration), Mon. Not. R. Astron. Soc. 469 (2017) 906
[38] A. Albert et al. (ANTARES Collaboration), Mon. Not. R. Astron. Soc. 500 (2021) 5614
[39] M. G. Aartsen et al. (IceCube Collaboration), Astrophys. J. 843 (2017) 112
[40] A. Albert et al. (ANTARES Collaboration), JCAP 03 (2021) 092
[41] J. Vandenbroucke et al. (IceCube Collaboration), ATel 12395 (2019)
[42] C. Mascaretti & F. Vissani, JCAP 08 (2019) 004
[43] S. Adrian-Martinez et al. (KM3NeT Collaboration), J. Phys. G: Nucl. Part. Phys. 43 (2016) 084001

Gravitational Waves and Black Holes

Fulvio Ricci[1,2,*]

[1]Sapienza Università di Roma, Roma Italy
[2]INFN Sezione di Roma, Roma, Italy

Abstract. We summarise the present status of the gravitational wave detectors on the Earth.Then, we discuss some of the most intriguing results obtained during the last data collection O3. We will focus on the statistical distributionof the population of these compact objects and we will discuss a couple of potential tests to be carried on the on the Bekenstein-Hawking thermodynamics using the gravitational wave signals.

1 Introduction

The experimental effort to detect Gravitational Waves (GW) was accomplished in 2015, just 100 years after the first presentation of the Einstein theory of General Relativity. To date, the catalog GWTC-3 of GW signals account for 90 events, the majority of them due to emission of the coalescence of binary black holes (BBH) systems. These are stellar mass black holes (BH) emitting GW in the acoustic frequency band of the LIGO, Virgo and KAGRA detectors. The signals have been collected during three observation runs, the last of which was O3 ended on March 27th, 2020. Towards the end of O3, the KAGRA detector, located in the Mozumi mine of the Kamioka mining of Japan, joined the observing run. This detector continued to operate for other two weeks taking data in coincidence with the German-UK gravitational-wave detector GEO600 near Hannover in Germany. The results from the GEO600-KAGRA run have been published in [1]. KAGRA, the two LIGOs, installed one in the Louisiana and the other in the Washington state, as Virgo, installed in Tuscany (Italy), are currently being upgraded fin preparation of the fourth observing run (O4). The joined data collection of the whole terrestrial network is expected to begin in 2023. We expecte to deal with detectors of higher sensitivity, so that the observation rate of gravitational waves will increase up to three times as often as in O3. The prediction is to have as many as five signals each week. The plans for upgrades in LIGO and Virgo concern different parts of the detectors. Virgo will modify the optical configuration adding the signal recycling mirror at the detector output, a solution already adopted in LIGO during the previous observation runs. Both US and European detectors will run using new high power laser and injecting trough the output port of the interferometer a squeezed electromagnetic (e.m.) vacuum with a frequency dependency. This is a significant novelty compared to the set-ups of LIGO and Virgo during O3. During the last run a phase squeezed vacuum field was injected in the interferometers, obtaining a sensitivity improvement at high frequency, but paying the price of a degrades at low frequency due to the anti-squeezing effect. The simultaneous reduction of the shot noise at high frequencies and quantum radiation pressure noise at low frequencies will be obtained by filtering the e.m.

*e-mail: fulvio.ricci@uniroma1.it

vacuum by means of a long Fabry-Perot cavity with low optical losses before to inject it in the output port of the interferometer. The filter cavity acts on the e.m. field like a passive filter in an electric circuit and it rotates the squeezed quadrature of the complex vector representing the e.m. vacuum in the Gauss plane as a function of frequency (see fig. **??**).

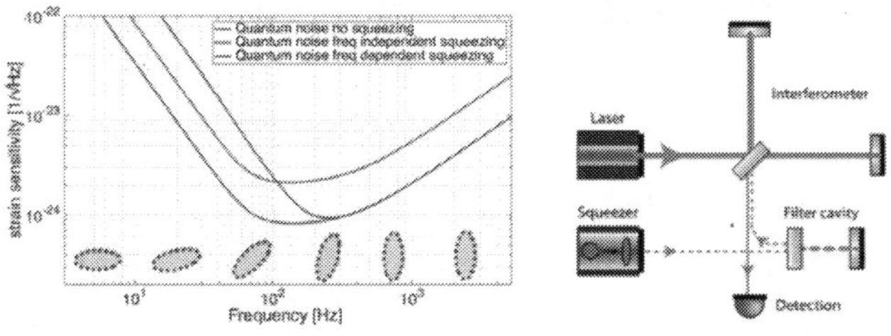

Figure 1. On the right we show a simplified optical scheme, which includes the cavity filtering the e.m. vacuum field (dashed line). On the left the sensitivity curves of the interferometer in he three different configurations: no squeezing, frequency independent and frequency dependent squeezing. On the bottom of the rotation of the vacuum squeezed quadrature in the Gauss plane as a function of frequency.

2 Population studies

The LIGO-Virgo collaborations published companion studies of the underlying population of black holes and neutron stars [2] and the history of the expansion of the Universe[3]. Moreover, the new catalog contains some surprises, such as an unusual neutron-star (NS) merged with a black-hole (BH), a massive BH merger, and BBHs revealing information about their spins. On the base of the 76 compact binary mergers detected with gravitational waves below a false alarm rate of 1 per year, present in the GWTC-3 catalog the LIGO-Virgo collaboration inferred some of the properties of the population of the astrophysical compact objects. The catalog contains three classes of binary mergers: BBH, BNS, and NSBH mergers. The inference on the BNS merger rate is ranging between 13 $\mathrm{Gpc}^{-3}\mathrm{yr}^{-1}$ and 1900 $\mathrm{Gpc}^{-3}\mathrm{yr}^{-1}$ while the NSBH merger rate is between 7.4 $\mathrm{Gpc}^{-3}\,\mathrm{yr}^{-1}$ and 320 $\mathrm{Gpc}^{-3}\mathrm{yr}^{-1}$, assuming a constant rate density versus comoving volume and taking the union of 90% credible intervals for all the methods used in this analysis. Concerning the BBH merger , the rate interval is 17.3 $\mathrm{Gpc}^{-3}\,\mathrm{yr}^{-1}$ - 45 $\mathrm{Gpc}^{-3}\,\mathrm{yr}^{-1}$ at a redshift (z=0.2), accounting for the evolution of the merger rate to with redshift z

The neutron star mass distribution is broad and it is extended from $1.2^{+0.1}_{-0.2}M_\odot$ to $2.0^{+0.3}_{-0.2}M_\odot$, without evidence of an increase of the merger rate before 10 $M_\odot$. The BBH mass distribution is localised over- and under-densities relative to a power law distribution. While we find the mass distribution of a binary's more massive component strongly decreases as a function of primary mass, we observe no evidence of a strongly suppressed merger rate above $\sim 60M_\odot$. The rate of BBH mergers seems to increase with redshift at a rate proportional to $(1+z)^\kappa$ with $\kappa = 2.7^{+1.8}_{-1.9}$ for $z \lesssim 1$. Concerning the BH spins, half of spin magnitudes are below $\chi_i \simeq 0.26$; in addition we have evidence of an increase of spin magnitude in systems with more unequal mass ratio and even negative aligned spins in the population [2].

The events of the GWTC-3 catalog have been used also to estimate the current value of the Hubble parameter H_o[3]. The analysis is based on 47 events, providing the luminosity distances to the source. The the redshift associated to each event has been inferred in two modes: a) using the redshifted masses , b) using the galaxy catalog. In the first method based on the BBH redshifted masses we should infer simultaneously the source mass distribution and the Hubble parameter in function of the redshift $H(z)$. Then, assuming that the assigned mass distribution does not evolve with z, the result obtained for H_o is:

$$H_o = 68^{+12}_{-7} \quad \frac{km}{s\,Mpc} \tag{1}$$

The use of the galaxy catalog implies to marginalise, in the statistical sense, over the redshifts of each event's potential hosts. Then, making the assumption that the BBH population is fixed, they conclude that H_o is :

$$H_o = 68^{+8}_{-6} \quad \frac{km}{s\,Mpc} \tag{2}$$

Although the gain to the respect of the values obtained using just the data of the previous catalog GETC-2 is of the order of 42% outcome, we note that it depends significantly on the assumption done on he BBH mass distribution.

3 GR Tests and BH Thermodynamics

There are several good motivation to search for *new* physics beyond the Einstein theory of Gravitation, on which today we are founding our models of the Universe. The matter we know and that makes up all stars and galaxies only accounts for 5% of the content of the universe, while an other kind of matter of unknown origin , the dark matter, seems to complement the Universe mass for about 27 %. In addition, cosmology at present has no explanation for about 68 % of the energy in the Universe that comes in the form of dark energy. Universe is a mystery, but today we have a new way for studying it via the gravitational waves. The GW advanced detectors allows to perform tests of general relativity (GR) in the strong-field regime of gravity. The most recent ones of the LIGO-Virgo-KAGRA collaboration is based on fifteen BBH events with a false alarm rate of $\leq 10^{-3}$ year $^{-1}$ to which a NS-BH event has been added. The test is carried on by computing, for each event ,the power residuals resulting from the subtraction of the waveform best fit of signal to the real data and check that is compatible with the detector noise. To find evidence for possible deviations from GR, we can perform *consistency* or *parameterised* tests. Consistency tests has been done by checking how much the signal shape is consistent with GR; the other category of tests implies the choice of an appropriate parametrisation to search for deviations from GR. Tests based on the post-Newtonian (PN) coefficients are sensitive to the physical effects that appear at different PN orders, while the parameterisation of the dispersion relation can give hints on non-GR wave propagation effects. The list of the GR tests is long and we refer to the article [4] for the details. Here we cite just the results, all null at present:

- the BH quadrupole moments induced by its rotation are consistent with those of Kerr black holes in GR;
- no evidence of dispersion of gravitational waves was found;
- no polarisation modes from GR prediction were evidenced;
- in the events that were analysed, no post-merger echoes were found.

In addition, we report the new upper bound on the graviton mass m_G set at 90% credibility:

$$m_G \leq 1.27 \cdot 10^{-23} \quad \frac{eV}{c^2} \tag{3}$$

GR is classified as a classical theory, leading to many surprising results as the prediction of the BH existence and its singularity inside its horizon. BH define a space volume from which even light can not escape, but we may expect that classical physics breaks down near the singularity. In addition there are good reasons to believe that at least in the proximity of the horizon we could observe deviations from semiclassical physics predictions [5],[6]. In a pure scenario depicted by GR, BH do not emit any radiation, so its absolute temperature is zero. However, in the 70s, S. Hawking and J. Bekenstein [7],[8],[9] theorised that even in a semiclassical framework, including some how quantum mechanics (QM) in the BH description, the compact object should acquires a finite temperature. The first hint that it could exist a connection between BH and Thermodynamics came is dated 1970: it concerns the mathematical result that the value of the BH horizon area always increases when additional matter or radiation falls into the black hole. Moreover, in the case of BBH merger the area of the horizon are of the final BH should be greater than the sum of the areas of the two primary BHs. These properties suggest that there is a resemblance between the BH horizon area and the thermodynamical concept of Entropy. Further, the similarity is extended looking at the change of BH internal energy, which pushes us to assume that a finite temperature of a BH is proportional to its surface gravity. The existence of a finite temperature for BH implies that this object is in thermal equilibrium with its external environment by emitting thermal radiation. Thus, we are ending up with the paradox that in the pure classical description of a BH, this object would absorb any thermal radiation that fell on it but by definition would not be able to emit anything in return. Al these considerations pushed to assume that the Bekenstein-Hawking entropy of a black hole depends on the BH area, the BH has a temperature and emits thermal radiation.
These are all intriguing theoretical consideration; now the GW experimental physics gives us access to a new regime in which the unique properties of BH can be observationally tested. Since the BBH merger is an irreversible process, we expect that the total horizon areas should always increase. The area is in turn a function of BH mass and spin and, using the values given by the GW catalog of events, it is rather an academic exercise to test the entropy increase for the coalescent events of binary black holes [10]. However, a fundamental objection has been raised on this approach to prove that we are really dealing with quantities related to BH thermodynamics: we are extracting data by fitting the GW data with numerical relativity formulas, i.e. in the full assumption of the validity of the classical GR theory. To evade this problem, it has been proposed [11] to avoid GR by using just the initial masses measured using the data of the quasi Keplerian orbit far from the merger phase and for the final BH mass the ring-down part of the signal.
An other way to approach the question of how to check the thermodynamical laws of BHs, is looking at the spin dependence. According to the Bekenstein-Hawking formula, the entropy is smaller in the case of larger spins for a given fixed mass M. The statistics interpretation of the entropy implies that there are fewer micro-states with large spin than with small spin. Thus, it is possible to compare spin data with the models predicting the BH population according to the micro-canonical ensemble. This would be a way to apply the statistical mechanics approach to the BH physics trough its entropy study: an interesting attempt is reported in [12].

The finite temperature of BH thermodynamics implies a thermal emission, i.e. a change of the BH mass and, as consequence, of its entropy. Assuming a statistical point of view, we can state that there is a change in the number of internal configurations accessible to the system or conversely, an amount of information emitted by the system. Bekenstein derived an universal bound for the maximum average rate of information emission achievable by a physical system[13]. Then, Hod noted that this bound applies to the ring-down phase of a BH relaxing towards equilibrium. He quantified the bound by introducing the information loss parameter $\mathcal{H}$, a quantity having an explicit dependence on the relaxation time of a perturbed black hole $\tau = 1/\omega_I$ [14]:

$$\mathcal{H}_{max} = \frac{1}{\pi}\frac{\hbar\omega_I}{k_B T} \leq 1 \quad (4)$$

where T is the BH temperature and k_B the Boltzmann constant. This limit is known as the Bekenstein-Hod universal bound and, using a time-domain ringdown analysis of BBH events of GW catalog of signals, Carullo et al. [15] were able to provide an experimental confirmation of the Bekenstein-Hod bound. This can be regarded as the first experimental verification of a long standing prediction on the dynamical information-emission process of a BH.

4 Conclusions

The accumulation of GW data, produced in the merger phase of astrophysical compact object as BH and NS, increases our chance to shed lights on the dark side of the Universe. 90 signals have been collected in the GWTC-3 and on this base more stringent values on the formation rates of BBH and BHNS have been derived. There is a growing theoretical and experimental evidence that modifications of GR at small and large energies are somehow inevitable and the GW observations allow to perform GR tests in the strong-field regime of gravity. Although no deviations have been observed up to now, significant upper bounds have been set on the deviation of post-Newtonian coefficients of different PN orders and on the parameters giving hints on non-GR wave propagation effects. New data will come soon with the upgraded LIGO-Virgo-KAGRA network that will begin the O4 observation in 2023. The goal is to run the network at a signal to noise ratio higher than in the past, increasing the chance to insight into the origins of the binary compact objects, the extreme conditions inside neutron stars and perform most-stringent tests of GR. We are also confident to detect new categories of GW signals, as those emitted by the asymmetric rotating neutron stars, giving hints on the status of central matter of the neutron star with densities exceeding that of atomic nuclei.

References

[1] Abbott, R. et al *First joint observation by the underground gravitational-wave detector KAGRA with GEO 600*, arXiv:2203.01270LIGO-P2100286

[2] Abbott R. et al. *The population of merging compact binaries inferred using gravitational waves through GWTC-3*, arXiv:2111.03634v4

[3] Abbott R. et al. *Constraints on the cosmic expansion history from GWTC-3* arXiv:2111.03604

[4] Abbott R. et al. *Tests of general relativity with GWTC-3* arXiv:2112.06861, *accepted for publication in Phys. Rev. D*

[5] Almheiri A., Marolf D. , Polchinski J., Sully J., *Black Holes: Complementarity or Firewalls?* J. High Energy Phys. **02**, 062, (2013).

[6] Hod S., *Universal bound on dynamical relaxation times and black-hole quasinormal ringing* Physical Review D **84**, 124030 (2011)

[7] Hawking, S.W. *Black holes and thermodynamics* Physical Review D **13**, 191 (1976)

[8] Bekenstein, J. D. *Black Holes and the Second Law.* Lettere al Nuovo Cimento **4**, 737?740 (1972)

[9] Bekenstein, J. D. *Black Holes and Entropy* Physical Review D **7**, 2333 (1973)

[10] Salerno, V. *Il segnale di onde gravitazionali GW150914 e la variazione di entropia associata* Bachelor's Thesis, University of Rome, Rome, Italy, 2022.

[11] Cabero M., Collin M., Capano D., Fischer-Birnholtz O., Krishnan B., Nielsen,A.B., Nitz A.H., Biwer C.M, *Observational tests of the black hole area increase law*, Physical Review D **97**, 124069 (2018)

[12] Bianchi E., Gupta A., Haggard H.M., Sathyaprakash B.S., *Small Spins of Primordial Black Holes from Random Geometries: Bekenstein-Hawking Entropy and Gravitational Wave Observations* arXiv 2018, arXiv:1812.05127v2.

[13] Bekenstein J. D., *Energy Cost of Information Transfer* Physical Review Letters **46**, 623 (1981).

[14] Hod S., *Universal bound on dynamical relaxation times and black-hole quasinormal ringing* Physical Review D **75**, 064013 (2007),

[15] Carullo G., Laghi D., Veitch J.,Del Pozzo W., Bekenstein-Hod Universal Bound on Information Emission Rate Is Obeyed by LIGO-Virgo Binary Black Hole Remnants Physical Review Letters **126**, 161102 (2021)

Very high energy sky monitoring with the Southern Wide-field Gamma-ray Observatory

Giovanni La Mura[1,*], *Ulisses* Barres de Almeida[2] , and *Francesco* Longo[3,4,5]

[1]Laboratório de Instrumentação e Física Experimental de Partículas (LIP), Avenida Prof. Gama Pinto 2, 1649-003 Lisboa, Portugal
[2]Brazilian Center for Physics Research (CBPF), Rua Dr. Xavier Sigaud 150, 22290-180 Rio de Janeiro, Brazil
[3]IFPU - Institute for Fundamental Physics of the Universe, Via Beirut 2, I-34014 Trieste, Italy
[4]INFN, Sezione di Trieste, via A. Valerio 2, I-34100 Trieste, Italy
[5]Dipartimento di Fisica, Università degli Studi di Trieste, via A. Valerio 2, I-34100 Trieste, Italy

Abstract. The Southern Wide-field Gamma-ray Observatory (SWGO) is the proposal for a new ground-based γ-ray instrument in the Southern Hemisphere, which will use an array of water-Cherenkov particle detectors to provide continuous monitoring of a large portion of the sky at the very- and ultra-high-energies (VHE and UHE, respectively). At the low energy side, SWGO aims to push the observational range of wide-field ground-based γ-ray facilities down to a few hundred GeV, thus bridging the gap between space and ground-based facilities in the monitoring of the VHE sky. In the high energy domain, on the contrary, it will benefit from the optimal coverage of the Galactic Plane to map the distribution of UHE sources in the inner parts of the Galactic disk and close to the Galactic Center, leading to an extraordinary improvement in our ability to identify their most likely counterparts. In this contribution, we describe the concept of SWGO and its potential to constrain the physics of VHE emission and particle acceleration in γ-ray sources powered by relativistic jets and energetic shocks. We finally discuss its role within the global network of multi-messenger facilities.

1 Introduction

High Energy Astrophysics has recently witnessed several ground breaking discoveries. After the first detection of gravitational waves (GW) by the LIGO/VIRGO Collaboration [1], the binary neutron star (NS) merger that originated GW 170817 was the first event observed in conjunction with a short Gamma-Ray Burst (GRB) and subsequently associated with a kilo-nova [2, 3], thus providing further insight in the physics of NS and GRB. In the meantime, monitoring observation programmes, carried out by the IceCube observatory [4] and the *Fermi*-Large Area Telescope mission (*Fermi*-LAT, [5]) led to the identification of the γ-ray flaring blazar TXS 0506+056 as the likely source of an ultra-relativistic neutrino [6]. These highly anticipated and, yet, eagerly awaited discoveries demonstrated that the observation of the Universe in γ rays and other energetic particles could probe the fundamental laws of Physics at scales that are far beyond the reach of any laboratory experiment.

*e-mail: glamura@lip.pt

In the Very High Energy domain (VHE, $E \geq 100\,\mathrm{GeV}$), the expected fluxes of incoming signals are very low and they require instrumented areas that cannot be supported by artificial satellites. However, primary particles with these energies can interact with the Earth's atmosphere and start extensive air showers (EAS) of photons and relativistic charged particles that propagate toward the ground and can eventually be detected. Their observation is based on two complementary techniques. The first one uses Imaging Atmospheric Cherenkov Telescopes (IACT) to observe the radiation flashes produced by the relativistic charged particles of the shower. The second one, instead, adopts arrays of particle detectors located in high-altitude sites to interact directly with the shower particles. IACT observatories, such as the Major Atmospheric Gamma-ray Imaging Cherenkov (MAGIC, [7]), the Very Energetic Radiation Imaging Telescope Array System (VERITAS, [8]), the High Energy Stereoscopic System (H.E.S.S., [9]) and the next-generation Cherenkov Telescope Array (CTA, [10]) offer the best performance in terms of sensitivity to point-like sources, spatial resolution and energy resolution. On the other hand, they have a narrow field of view (FoV), up to a few square degrees, and they can only operate in dark conditions, making them unsuitable for continuous monitoring and extensive survey programmes. The particle detector arrays, such as the Astrophysical Radiation with Ground-based Observatory (ARGO, [11]), the Tibet Air Shower experiment (Tibet ASγ, [12]) the High Altitude Water Cherenkov (HAWC, [13]), and the recently completed Large High Altitude Air Shower Observatory (LHAASO, [14]), conversely, provide simultaneous coverage of large sky areas, up to approximately 2 sr, and they can operate almost continuously. Although we have a well-established strategy to grant nearly all-sky access to IACT facilities, at present, VHE γ-ray monitoring observatories are only being operated in the Northern hemisphere.

The Southern Wide-field Gamma-ray Observatory project (SWGO, [15]) aims at designing the latest generation, highly efficient and cost effective solution to install an extensive air-shower (EAS) detector array in the Southern hemisphere. Here we describe the fundamental concepts of the project, we discuss its scientific potential and we compare the performance estimates with the known or expected properties of possible targets. Finally, we conclude the contribution by discussing the role of SWGO in the framework of the global Multi-Messenger Astronomy network.

2 Project overview

The SWGO Collaboration[1] is currently formed by more than 50 institutions from 14 member countries. Additional support by associated scientists allows the involvement of the international community from even more countries. The primary goal of the collaboration is to produce a fully detailed solution to install a next-generation EAS array in the Southern hemisphere. Taking advantage of the continuous scanning of a wide FoV, such an observatory would complement the Northern facilities, attaining a nearly all-sky monitoring footprint, as illustrated in Figure 1.

The detector units used to build the array will be based on Water Cherenkov Detectors (WCD), which will be designed to optimize the signal collection and the particle recognition capabilities, thanks to very fine time resolution ($\Delta t \approx 2\,\mathrm{ns}$) and signal reconstruction techniques. The detector units may be deployed in an array consisting of a compact core, with a high filling factor ($ff \sim 80\%$) and an area of $S \approx 80000\,\mathrm{m}^2$, to track the low-energy showers initiated by sub-TeV primary particles, surrounded by a sparse array, with a filling factor degrading down to 1%, but extending to the order of $1\,\mathrm{km}^2$, which will instead observe the showers produced by the highest energy primaries.

[1] https://www.swgo.org

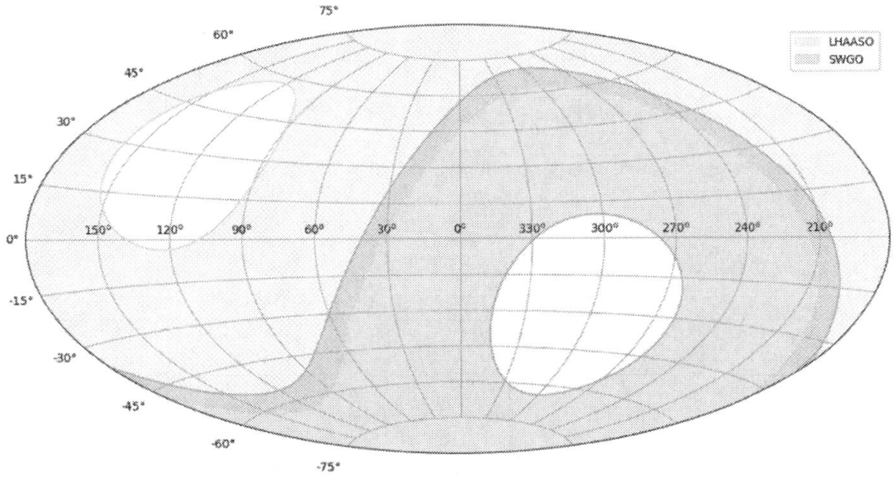

Figure 1. Representation of the visible sky, within 30° from zenith, showing the sky regions covered by LHAASO (cyan shaded area) and by SWGO, assuming an observatory latitude close to 23° S (orange shaded area). The map is plotted in Galactic coordinates.

The scientific potential of this experiment is huge. On one side, its location in the Southern hemisphere grants an optimal visibility of sky regions that cannot be currently monitored and that are very likely hosting a wealth of energetic sources, yet to be discovered. On the other, the experience gained with currently available monitoring facilities gives us an invaluable opportunity to design an instrument that can operate as a driver for follow-up campaigns, both by mapping the distribution of targets and by extending the simultaneous sky coverage of VHE monitoring facilities, in case of occurrence of transient and multi-messenger triggers. Its location close to the Southern Tropic, moreover, will result in an optimal coverage of the declination band of VHE transients that may be associated with IceCube and KM3NeT neutrino alerts. The project is presently in its Research and Development state (R&D), with the estimate of providing a fully detailed design proposal by the end of the year 2023. Construction and commissioning phases aim at completing the observatory in the 2030s, with a planned experiment duration of ten years.

3 SWGO Science goals

SWGO will open a new window to approach some of the most compelling problems in today's high energy particle physics and astrophysics, ranging from the nature of high-energy emission in extra-galactic transients to the origin of PeV cosmic rays (CR) and the search for dark matter particles. In many cases, it will represent an unrivalled facility for the exploration of time-variability aspects, as well as for mapping extended sources, such as the *Fermi Bubbles* and the star-forming regions where recent super-nova activity is expected to power *super-bubbles*. In addition, its extended energy coverage is intended to make it able to issue alerts for follow-up campaigns, thus complementing the global multi-messenger network of gravitational, electromagnetic, cosmic-ray and neutrino observatories. General descriptions of the scientific questions that SWGO can address have been provided in the form of a white paper [16]. Here we discuss a set of driving Science cases, also summarized in Table 1, for which recent studies and new findings, based on the *Fermi*, HAWC, and LHAASO

Table 1. SWGO Science driving benchmarks.

Target	Design driver	Expected requirement
Transient sources Gamma-Ray Bursts	Low-Energy sensitivity and site altitude	Minimum integration for 5σ detection: $F(E > 100\,\mathrm{GeV}) \approx 10^{-8}\mathrm{erg\,cm^{-2}\,s^{-1}}$ $N(E,t) \propto E^{-2} \cdot t^{-1.2}$
Galactic Accelerators: PeVatron Sources	High-energy sensitivity and energy resolution	Maximum exp-cutoff energy detectable at 95% CL in 5 years for: $F(1\,\mathrm{TeV}) = 5\,\mathrm{mCrab}$, $N(E) \propto E^{-2.3}$
Galactic Accelerators: PWNe and TeV Halos	Extended source sensitivity and angular resolution	Maximum source angular extension detectable at 5σ in 5-yr integration for $F(> 1\,\mathrm{TeV}) = 5 \times 10^{-13}\,\mathrm{TeV\,cm^{-2}\,s^{-1}}$
Diffuse Emission: Fermi Bubbles	Background rejection	Cosmic-Ray residual background level $< 10^{-5}$ at 1 TeV
Fundamental Physics: Dark Matter from Galactic Halo	Mid-range energy sensitivity and site latitude	Maximum energy for $b\bar{b}$ thermal-relic cross-section limit at 95% CL for Einasto profile
Cosmic-rays: Mass-resolved dipole/ multipole anisotropy	Muon counting capability	Maximum dipole energy at 10^{-3} level log-mass dipole resolution at 1 PeV - goal is to achive $A = \{1, 4, 14, 56\}$; Maximum multipole scale > 0.1 PeV

experience, revealed the properties of different possible targets, suggesting what design characteristics will be needed to set the instrument's discovery potential.

3.1 Extra-galactic and transient sources

At the lowest energies, the core science of SWGO is focused on transient sources, exploiting the wide-field of view and near-continuous duty cycle of the observatory, along with its southern location, to work as a monitoring and trigger instrument complementary to CTA. At the few-hundred GeV energy scale, the two principal target sources are Active Galactic Nuclei (AGN) and GRB, which are, respectively, the candidate multi-messenger astrophysical counterparts of ultra high energy (UHE) neutrinos and GW events [2, 3, 6], and allow to probe the spectrum and intensity of the Extragalactic Background Light (EBL), through its redshift dependent opacity effects [17–19]. In terms of performance goals, the aim is to guarantee that SWGO will be an effective detector of serendipitous transients, able to issue fast triggers for other instruments, particularly in the case of GRBs, where it should have good performance in the first few kiloseconds from t_0, when source flux is higher [20].

As an example, Figure 2 presents an analysis of the detectability of the bright GRB 130427A at VHE energies, based on a light-curve extrapolation into the VHE domain [21], using different instrument performances and energy threshold prescriptions. The calculations show that an integral target sensitivity $\sim 10^{-9}\mathrm{ph\,cm^{-2}\,s^{-1}}$ between $0.1 - 1\,\mathrm{TeV}$ (in 1 ks) is able to detect a source only a fraction of the brightness of GRB 130427A. This level roughly corresponds to the 1/6 brightest GRBs in the *Fermi*-LAT catalogue, and is statistically equivalent to securing one safe serendipitous GRB detection by SWGO per year, being additionally sufficient to guarantee regular (monthly) AGN flare triggers [22]. For bright and relatively nearby GRBs, such as GRB 190114C [23], this sensitivity would allow not only for the generation of an early (< 1 ks) trigger alert by SWGO, but also, in case of favourable observing conditions, of a short-time resolution of the early emission below the 100 s time-scale [21], with the possibility of probing the prompt phase in the VHE spectral range.

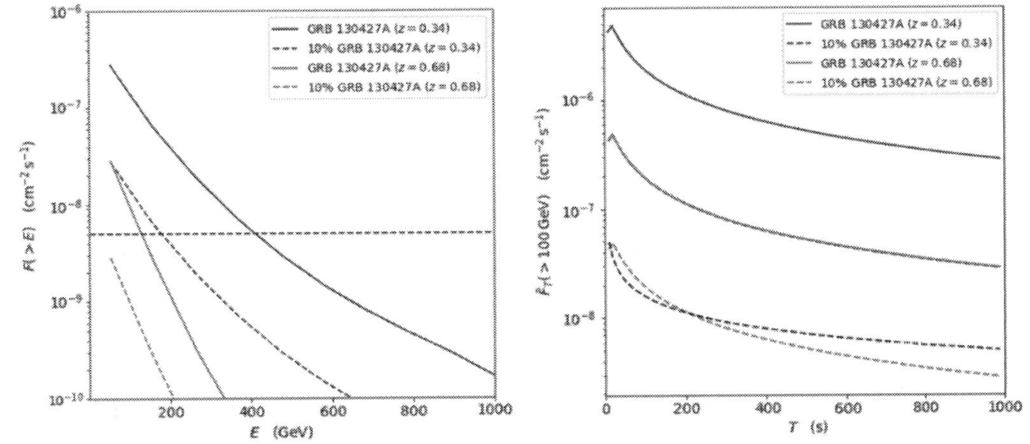

Figure 2. Left panel: average photon flux expected above any given energy in 1000 s for a burst with the spectral and temporal characteristics of GRB 130427A (blue continuous line), an identical burst with 10% its strength (blue dashed line) and the same previous cases computed for twice the measured redshift (red continuous and dashed lines). The horizontal black dashed line represents a reference flux limit of $5 \cdot 10^{-9}$ ph cm^{-2} s^{-1}. Right panel: same as left panel, but calculated for the instantaneous integral photon fluxes, expected above 100 GeV and assuming that the limiting flux for a detection scales approximately as the square root of the observation time.

3.2 Ultra-high energy domain

At the high extreme of SWGO's energy range, science is dominated by the search for PeVatrons. These are the putative sources responsible for the acceleration of knee cosmic-ray particles, whose signature in γ rays is expected to come in the form of a hard spectrum, with unabated emission beyond 100 TeV [24]. This science case has recently been reinforced by the Tibet ASγ detection of diffuse γ-ray emission in the Galaxy well above this energy range [25] and by LHAASO's detection of a number of PeV-emitting γ-ray sources in the Galactic Plane [26]. However, differences in the observing pattern of Tibet AS+MD and LHAASO result in a problematic reconstruction of the diffuse component, implying the need to cover a larger portion of the Galactic Plane. The identification of PeVatrons by means of the detection of a high-energy spectral cutoff signature, on the other hand, poses a requirement on the energy resolution of the observatory, of $\sim$ 30% in the energy range above 100 TeV. A firm estimate on the required point-source sensitivity at these energies depends on knowledge of the Galactic hard-spectrum sources, which is largely missing. Still recent estimates suggest that an integral sensitivity above 100 TeV at the 10^{-13} ph cm^{-2} s^{-1} level (for 1-year integration) is required for an in-depth probe of the Galactic PeVatron population [24], where the lack of sources with spectral index close to -2 in LHAASO data is a matter of concern that may be solved by covering the central regions of the Galaxy [26].

The study of extended sources and of diffuse emission, whose highest-energy flux is generally linked to the fraction of PeVatrons among the Galactic accelerators, will critically depend on the background rejection power of the array, which aims to achieve a residual charged cosmic-ray background better than 10^{-3} at 10 TeV, and beyond $\sim 10^{-4}$ above 100 TeV. Above $\sim$ 10 TeV, the γ/hadron separation is strongly associated with the capability to detect single muons at the individual detector units. The science case for extended sources will be centered on the study of Pulsar Wind Nebulae (PWNe) and TeV Halos, which pose the principal

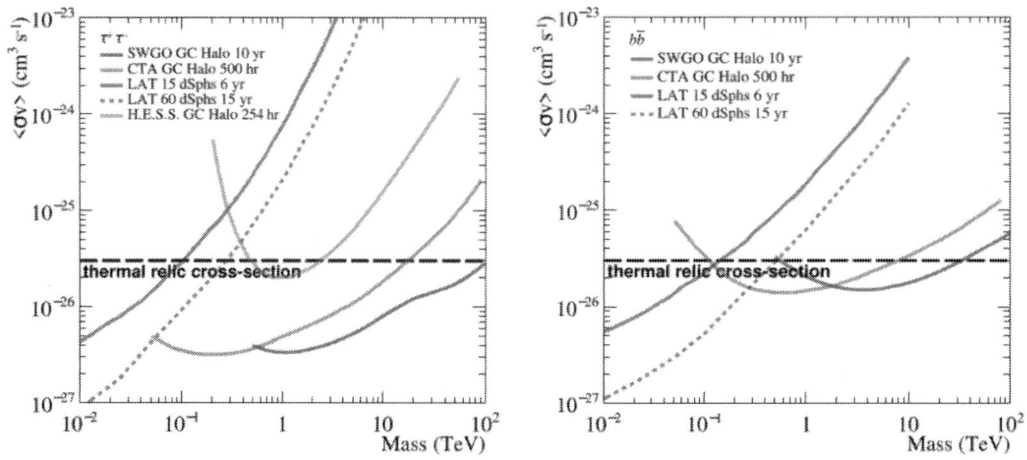

Figure 3. Expected sensitivity of SWGO to the detection of thermal relic Dark Matter signals from the Galactic Center Halo for the channels $\tau^+\tau^-$ (left) and $b\bar{b}$ (right). The plots show that the expected sensitivity of SWGO should allow to probe deep into the entire range of the thermal relic cross-section limits, extending CTA capabilities above a few TeV, up to 100 TeV. Figure adapted from [28].

benchmark constraints for the angular resolution of the observatory. Here, the goal will be to extend the detectability of these objects towards higher energies, above 10 TeV, and for angular extensions up to $\sim 3°$, for which a relatively good angular resolution capability of approximately 0.15° is estimated to be necessary to resolve the vast majority ($> 97\%$) of the population [27].

3.3 Dark matter searches

The sensitivity and energy resolution of SWGO should also be sufficient to constrain the science goals for Dark Matter (DM) observations, for which the Galactic Center Halo is the most promising observational target, followed by some potentially interesting ultra-faint Dwarf Spheroidal galaxies (DSphs) that could be detected, for instance, by the Large Synoptic Survey Telescope (LSST, [29]). As shown in Figure 3, the best contribution of SWGO in the study of DM is its possibility to constrain the entire energy range of weakly interacting particle models (WIMP), especially above 30 TeV, where SWGO sensitivity could outperform that of CTA for reasonable integration times. In order to obtain this result, a point-source flux sensitivity of $\approx 3 \times 10^{-14}$erg cm^{-2} s^{-1} in the range between a few to several tens of TeV (for 5-year integration), as used in the figure, is the necessary requirement.

3.4 Cosmic-ray studies

At high energies, generally above $\sim$ 10 TeV, the capability to detect single muons is crucial for a good γ/hadron separation. This capability would have a large impact on the background rejection necessary for improving detectability of faint diffuse γ-ray emission [30], but it would also allow for unique mass-resolved charged cosmic-ray studies up to PeV energies. The science goal here is ultimately to understand the nature and the evolution of the cosmic-ray dipole and multipoles between 0.1 and several PeVs. The CR anisotropy investigation requires extended sky coverage, which is crucial for multipoles with low l-scale, but may be useful to search for variations in different directions for large l-scale studies, as well. The capabilities of SWGO aim at extending the anisotropy detection beyond the PeV-scale, investigating the CR spectrum in four mass groups ($A = \{1, 4, 14, 56\}$) and mesuring the multipole l-scale beyond 0.1 PeV.

4 Conclusions and outlook

Since the first VHE γ-ray detection [31], hundreds of sources have been discovered, including variable objects. Their study requires instruments able to continuously monitor large portions of the sky, sensitive to energies above those of satellite-based experiments. The detection of photons of extreme energies from the Galaxy [25, 26] strengthens the necessity for a large survey instrument with optimal access to the Galactic Plane and sensitivity beyond 100 TeV. In addition, the VHE properties observed in the extremely bright GRB 221009A [32–35], which has been observed to emit photons up to 18 TeV by LHAASO [36] from a redshift of $z = 0.151$ [37], further outlined the importance of obtaining all-sky coverage from VHE monitoring instruments. This will increase the number of GRB triggers that can be promptly covered and, consequently, probe their VHE properties and their interaction with the EBL with enhanced statistical significance.

The vision pursued for SWGO is of an observatory that aims to cover a wide energy range, bridging between satellite observations down to 100 GeV, and reaching towards the PeV energy domain [15]. Its location at a high altitude in the Southern hemisphere will provide a new window to an unexplored sector of the sky and, particularly, to the center of the Galaxy. Currently, in the R&D Phase, the project plans to deliver an Observatory proposal by 2023.

Acknowledgements

The SWGO Collaboration acknowledges the support from the agencies and organizations listed here: https://www.swgo.org/SWGOWiki/doku.php?id=acknowledgements.

References

[1] B.P. Abbott, *et al.* (LIGO Scientific Collaboration and Virgo Collaboration), Phys. Rev. Lett. **116**, 061102 (2016)
[2] B.P. Abbott, *et al.*, The Astrophysical Journal Letters **848**, L12 (2017), 1710.05833
[3] B.P. Abbott, *et al.*, The Astrophysical Journal Letters **848**, L13 (2017), 1710.05834
[4] The IceCube Collaboration, Journal of Instrumentation **12**, P03012 (2017), 1612.05093
[5] W.B. Atwood, *et al.*, The Astrophysical Journal **697**, 1071 (2009), 0902.1089
[6] The IceCube Collaboration, Science **361**, eaat1378 (2018), 1807.08816
[7] J. Aleksić, *et al.*, Astroparticle Physics **72**, 61 (2016), 1409.6073
[8] T.C. Weekes, *et al.*, Astroparticle Physics **17**, 221 (2002), astro-ph/0108478
[9] F. Aharonian, *et al.*, Astronomy & Astrophysics **457**, 899 (2006), astro-ph/0607333
[10] Cherenkov Telescope Array Consortium, *Science with the Cherenkov Telescope Array* (World Scientific Ltd, 2019)
[11] A. Aloisio, *et al.*, Nuovo Cimento C Geophysics Space Physics C **024**, 739 (2001)
[12] M. Amenomori, *et al.*, Physical Review Letters **69**, 2468 (1992)
[13] T. DeYoung, Nuclear Instruments and Methods in Physics Research A **692**, 72 (2012)

[14] Z. Cao, the LHAASO Collaboration, Chinese Astronomy and Astrophysics **43**, 457 (2019)

[15] J. Hinton, SWGO Collaboration, *The Southern Wide-field Gamma-ray Observatory: Status and Prospects*, in *37th International Cosmic Ray Conference. 12-23 July 2021. Berlin* (2022), p. 23, 2111.13158

[16] A. Albert, *et al.*, arXiv e-prints arXiv:1902.08429 (2019), 1902.08429

[17] E. Dwek, F. Krennrich, Astroparticle Physics **43**, 112 (2013), 1209.4661

[18] A. Franceschini, L. Foffano, E. Prandini, F. Tavecchio, Astronomy & Astrophysics **629**, A2 (2019), 1907.03666

[19] A. Saldana-Lopez, A. Domínguez, P.G. Pérez-González, J. Finke, M. Ajello, J.R. Primack, V.S. Paliya, A. Desai, Monthly Notices of the Royal Astronomical Society **507**, 5144 (2021), 2012.03035

[20] M. Ajello, *et al.*, The Astrophysical Journal **878**, 52 (2019), 1906.11403

[21] G. La Mura, U. Barres de Almeida, R. Conceição, A. De Angelis, F. Longo, M. Pimenta, E. Prandini, E. Ruiz-Velasco, B. Tomé, Monthly Notices of the Royal Astronomical Society **508**, 671 (2021), 2109.06676

[22] G. La Mura, G. Chiaro, R. Conceição, A. De Angelis, M. Pimenta, B. Tomé, Monthly Notices of the Royal Astronomical Society **497**, 3142 (2020), 2001.04503

[23] MAGIC Collaboration, Nature **575**, 455 (2019), 2006.07249

[24] E.O. Angüner, F. Cassol, H. Costantini, C. Trichard, G. Verna, *Cherenkov Telescope Array potential in the search for Galactic PeVatrons*, in *36th International Cosmic Ray Conference (ICRC2019)* (2019), Vol. 36 of *International Cosmic Ray Conference*, p. 618, 1911.06134

[25] M. Amenomori, *et al.*, Physical Review Letters **126**, 141101 (2021), 2104.05181

[26] Z. Cao, *et al.*, Nature **594**, 33 (2021)

[27] R. Lopez-Coto, A. Mitchell, E.O. Anguner, G. Giacinti, SWGO Collaboration, *Galactic Science with the Southern Wide-field Gamma-ray Observatory*, in *37th International Cosmic Ray Conference. 12-23 July 2021. Berlin* (2022), p. 892, 2109.03521

[28] A. Viana, H. Schoorlemmer, A. Albert, V. de Souza, J.P. Harding, J. Hinton, Journal of Cosmology and Astroparticle Physics **2019**, 061 (2019), 1906.03353

[29] J.D. Simon, Annual Reviews in Astronomy & Astrophysics **57**, 375 (2019), 1901.05465

[30] A. Neronov, D. Semikoz, Phys. Rev. D **102**, 043025 (2020)

[31] T.C. Weekes, M.F. Cawley, D.J. Fegan, K.G. Gibbs, A.M. Hillas, P.W. Kowk, R.C. Lamb, D.A. Lewis, D. Macomb, N.A. Porter et al., The Astrophysical Journal **342**, 379 (1989)

[32] S. Dichiara, J. Gropp, J. Kennea, N. Kuin, A. Lien, F. Marshall, A. Tohuvavohu, M. Williams, the Neil Gehrels Swift Observatory Team, GRB Coordinates Network **32632**, 1 (2022)

[33] P. Veres, E. Burns, E. Bissaldi, L. Lesage, O. Roberts, on behalf of the *Fermi* GBM Team, GRB Coordinates Network **32636**, 1 (2022)

[34] S. Lesage, P. Veres, O. Roberts, E. Burns, E. Bissaldi, on behalf of the *Fermi* GBM Team, GRB Coordinates Network **32642**, 1 (2022)

[35] E. Bissaldi, N. Omodei, M. Kerr, on behalf of the *Fermi*-LAT Team, GRB Coordinates Network **32637**, 1 (2022)

[36] Y. Huang, H. Shicong, C. Songzhan, Z. Min, L. Cheng, Y. Zhiguo, C. Zhen, the LHAASO Collaboration, GRB Coordinates Network **32677**, 1 (2022)

[37] A. de Ugarte Postigo, *et al.*, GRB Coordinates Network **32648**, 1 (2022)

Attaining the PeV frontier of the cosmic ray spectrum in space with HERD

Chiara Perrina[1,*] on behalf of the HERD Collaboration

[1]Institute of Physics, Ecole Polytechnique Fédérale de Lausanne (EPFL), CH-1015 Lausanne, Switzerland

Abstract. The High Energy cosmic-Radiation Detection facility (HERD) is a calorimetric experiment planned to be launched in 2027. It will be operational for at least 10 years on board the China Space Station. With HERD we will measure the energy flux of cosmic protons and heavier nuclei from 30 GeV up to, for the first time in space, a few PeV. We will search for signatures of annihilation and decay products of dark matter in the energy spectrum of cosmic electrons and gamma rays from 10 GeV to 100 TeV. A wide field of view monitoring of the gamma-ray full-sky from 100 MeV will also be performed. The five HERD subdetectors, the calorimeter (CALO), the scintillating fiber tracker (FIT), the plastic scintillator detector (PSD), the silicon charge detector (SCD) and the transition radiation detector (TRD), are currently under development. The design, prospects and expected performance of HERD, as well as its contribution to the multimessenger astronomy will be presented in this contribution.

1 Introduction

- The most accurate values of the cosmological parameters, provided by the Planck mission [1], show that about 84% of the matter present in the universe consists of dark matter. Nevertheless, it has not yet been detected and its nature, origin and properties are still unknown. A method to detect the dark matter is the search for its annihilation and decay products in space, which may produce unique features in the cosmic electron (plus positron) and gamma-ray energy spectra.
- The indirect cosmic-ray measurements performed with ground-based experiments reveal a steepening in the all-particle spectrum between 1 PeV and 10 PeV. The origin of the so-called "knee" of the spectrum is still not clear. Assuming that cosmic rays below 10^{18} eV have Galactic origin, the knee could indicate that most cosmic accelerators in the Galaxy have reached their maximum energy to accelerate protons [2]. Direct measurements in space of the flux and composition of cosmic rays in the PeV region will shed light on the knee origin and the acceleration and propagation mechanisms of cosmic rays.
- Several new populations of extreme astrophysical objects were discovered by wide field of view (FOV) gamma-ray space telescopes in the GeV energy range, and by narrow FOV gamma-ray ground telescopes in the hundreds GeV range. In particular, the wide FOV

*e-mail: Chiara.Perrina@epfl.ch

space telescopes often provide crucial guidance on the observations of ground telescopes. The most powerful ground-based Cherenkov Telescope Array (CTA) [3] currently under development may not have the needed guidance once the Fermi-LAT satellite [4] will stop operations. A new wide FOV space gamma-ray telescope is therefore urgently needed.

Starting from 2027, the planned launch year, and for at least 10 years, HERD will be on board the China Space Station to address the above major problems in fundamental physics and astrophysics. Thanks to its unprecedented acceptance, HERD will open the opportunity to investigate cosmic rays at energies never attained before in space.

2 HERD physics observables and objectives

2.1 Cosmic Electrons and Positrons

HERD will perform high statistics precise measurements of the electron plus positron (CRE) flux up to about 100 TeV. The study of the CRE flux above 1 TeV is very interesting because electrons from 100 GeV lose energy via inverse Compton scattering and synchrotron radiation and therefore have a limited propagation distance. Electrons above few TeV are expected from local sources (< 1 kpc distance) or from dark matter annihilation and/or decay. Fig. 1 (left) shows the expected CRE as measured by HERD with five years of data. HERD will be able to detect spectral cutoff at high energy and local astrophysical sources. High precision measurements of AMS-02 [5] confirmed a rise from 8 GeV to 500 GeV in the positron fraction first revelead by PAMELA [6] that could be consistent with an astrophysical origin or a dark matter origin of positrons. The recent DAMPE result in the CRE spectrum shows a break at 0.9 TeV [7], that could be explained with the existence of an energy limit of the cosmic electron spectrum [8]. In that case, the cutoff of the positron spectrum should be below 0.9 TeV as seems to be favored by the AMS-02 measurements [9]. A way to clarify the origin of the positron excess is to measure the fine structure of the CRE spectrum with a high precision. As shown in Fig. 1 (right) HERD will clearly distinguish between a smooth spectrum, due to a pulsar wind nebula (PWN), or a spectrum with a "kink", due to dark matter with one year of data.

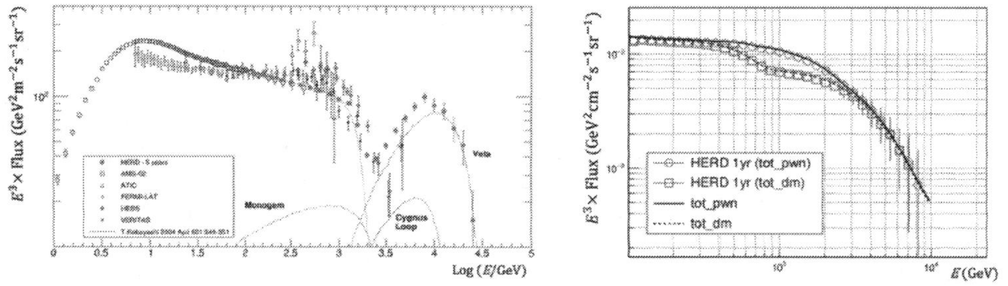

Figure 1: Expected $e^- + e^+$ spectrum as measured by HERD with five years (left) and one year (right) of data.

2.2 Cosmic nuclei

HERD will measure the flux of protons and helium nuclei up to a few PeV and of heavier nuclei (at least up to iron) up to a few hundreds of TeV/nucleon thus providing insights into the production, acceleration and propagation mechanisms of cosmic rays. HERD will explore directly in space the "knee" of the all-particle flux: the region where ground-based experiments observed a change of the spectral index, that could indicate the confinement limit of high energy particles by the galactic magnetic field, and the contribution of extragalactic sources. Fig. 2 shows the simulated proton (left) and helium (center) energy spectra and boron-to-carbon flux ratio (right) as measured by HERD with five years of data.

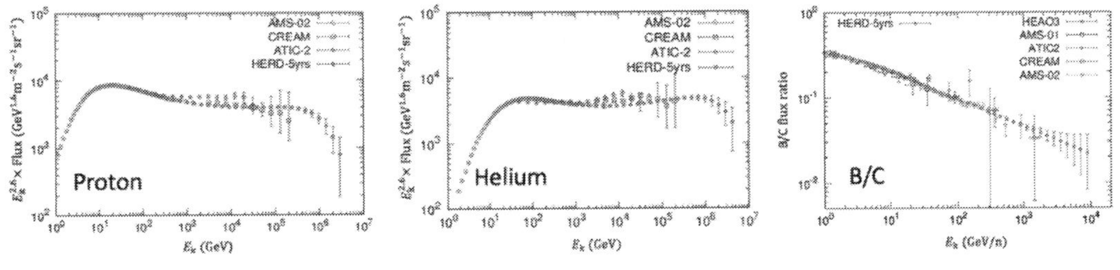

Figure 2: Expected proton (left) and helium (center) energy spectra and boron-to-carbon flux ratio (right) as measured by HERD with five years of data.

2.3 Cosmic gamma rays

Thanks to its large acceptance and sensitivity, HERD will be able to perform a gamma-ray full-sky survey from 100 MeV to study galactic and extragalactic gamma-ray point sources and diffuse emission, as well as the high energy gamma-ray transient phenomena. In the era of multimessenger and multiwavelength astronomy, it is essential to monitor the high energy gamma-ray sky in space, as was demonstrated by the Fermi-LAT detection [10] of the electromagnetic counterpart of the gravitational wave from a binary neutron star merger detected by LIGO. HERD with its unusual large field of view and energy coverage will play a unique role in multiwavelength studies across the electromagnetic spectrum, complementary to other space and ground telescopes measuring radio, optical, X-ray and gamma-ray signals. Its role will also be unique in the search for the electromagnetic counterpart of gravitational waves (detected by LIGO [11], Virgo [12], KAGRA [13]) and neutrino events (detected by IceCube [14], KM3NeT [15]). HERD will produce alerts for active galactic nuclei, novae and binary systems. The sky map expected from HERD after five years of activity is illustrated in Fig. 3. Another particularly interesting detection objective of HERD is the search for dark matter in the gamma-ray spectrum. Two-body dark matter annihilation or dark matter decay could be observed as a sharp peak over the high energy diffuse gamma-ray background. The gamma-ray line feature is considered as the "smoking gun" signal of dark matter. HERD will perform the gamma-ray line search from 10 GeV to 100 TeV with the best sensitivity ever achieved.

3 The HERD detector

The HERD detector (Fig. 4) consists of a 3D calorimeter (CALO), surrounded on the four lateral sides and on the top side by a sector of the scintillating-fiber tracker (FIT), the plastic scintillator detector (PSD), and the silicon charge detector (SCD); a transition radiation detector (TRD) is placed on one lateral side. With this 3D configuration, the total geometric factor of HERD is about 10 times larger than that of current-generation detectors. HERD will have a mass < 4000 kg, with an overall volume of $\sim 3.0 \times 2.3 \times 1.7$ m^3. The design of the HERD detector was also studied to reduce the systematic uncertainties (especially on the

 https://doi.org/10.1051/epjconf/202328001008

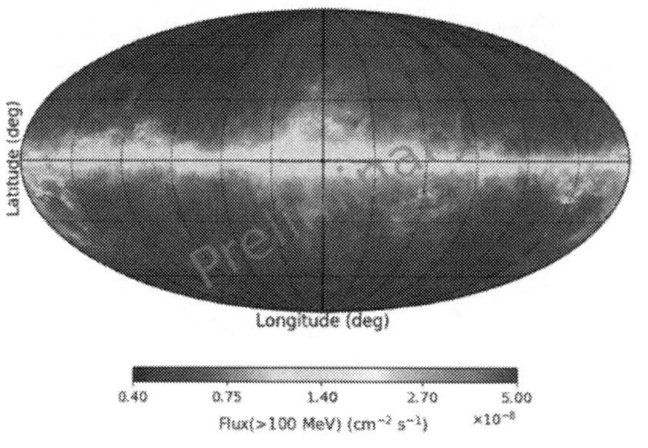

Figure 3: Expected map of the gamma-ray sky [16] in galactic coordinates with five years of HERD data.

absolute energy scale). For this purpose TRD calibrates the response of the calorimeter to high energy hadronic showers and each cube of CALO is read out by two independent systems: a IsCMOS camera and two photodiodes. SCD is the outermost detector to measure the particle charge before any possible nuclei fragmentation. The HERD expected performance in terms of energy range, energy, angular and charge resolution, electron/proton separation and geometric factors are listed in Table 1.

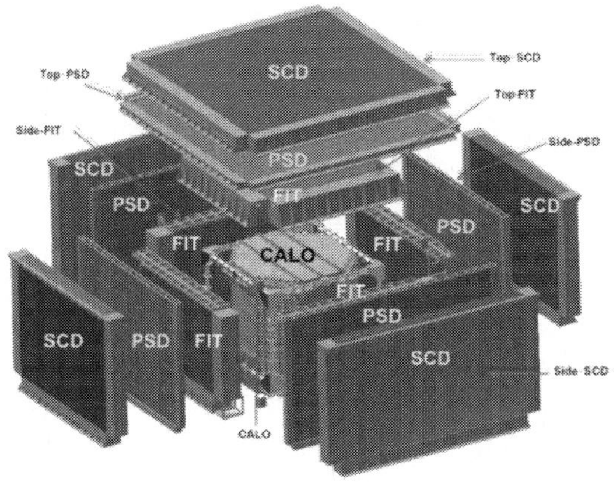

Figure 4: Exploded view of the HERD detector.

3.1 The calorimeter (CALO)

CALO [17] is a 3D homogeneous, isotropic and finely segmented calorimeter which accepts particles coming from every direction but the bottom side that is blocked by the device services and mechanical support. CALO is made of about 7500 LYSO cubes with edge length of 3 cm arranged in a sphere-like shape (Fig. 5) with a total depth of about 55 radiation lengths and 3 nuclear interaction lengths. The scintillation light induced by charged particles in the crystals is read out by two independent systems: the first one consists of wavelength shifting fibers (WLS) coupled to image Intensified scientific CMOS (IsCMOS) cameras, the second one is made of photodiodes [18] connected to custom front-end electronics chips called

Table 1: Expected HERD performance

Energy range (e/γ)	10 GeV - 100 TeV
Low energy range for γ	100 MeV - 10 GeV
Energy range (nuclei)	30 GeV - few PeV
Angular resolution (e/γ)	0.1° at 10 GeV
Charge resolution (nuclei)	10% - 15% for $Z = 1$ - 26
Energy resolution (e/γ)	< 1% at 200 GeV
Energy resolution (p)	20% at 100 GeV - PeV
e/p separation power	$> 10^6$
Field of view	± 90°
Geometric factor (e)	> 3 m^2sr at 200 GeV
Geometric factor (p)	> 2 m^2sr at 100 GeV

HIDRA. The double read-out system allows for redundancy, independent trigger, and cross calibration to reduce the systematic uncertainties.

Figure 5: Sketch of the sphere-like crystal array of the HERD calorimeter.

3.2 The scintillating-fiber tracker (FIT)

FIT [19] consists of five tracking sectors, made of 7 tracking planes, for 7 independent measurements of the position of a traversing charged particle (Fig. 6). Each tracking plane consists of two layers of FIT modules measuring the two orthogonal spatial coordinates. A module includes one scintillating-fiber mat and three silicon photomultiplier (SiPM) arrays to read out the scintillation light induced by the charged particles hitting the mat. The mat is made by stacking 6 layers of fibers for a width of 97.8 mm to match three Hamamatsu S13552-10 SiPM arrays. The fibers, manufactured by Kuraray (type SCSF-78MJ), have a round section with an average total diameter of 250 μm with a polystyrene core surrounded by two claddings. FIT aims at reconstructing the trajectories and the charge absolute value of charged cosmic rays, favor the conversion of low energy gamma rays and reconstruct the tracks of the produced electrons and positrons. Simulation studies are being conducted to optimize the tracker performance in terms of geometrical acceptance, angular resolution and converting capability while simultaneously being mechanically stable and complying with the thermal and power requirements set by the space station. In this configuration, FIT has 225792 read-out channels. So far the 128 channels of each SiPM array of FIT prototypes are read out by two VATA64 HDR16.2 ASIC chips made by IDEAS. A dedicated ASIC, called

Figure 6: Sketch of the HERD scintillating-fiber tracker (FIT).

BETA [20], with a larger dynamic range and a power consumption less than 1 mW/channel is being developed by the Institute of Cosmos Sciences of the University of Barcelona. BETA will have a dynamic range sufficient for charge measurements up to $Z = 26$.

3.3 The plastic scintillator detector (PSD)

PSD [21] will discriminate incident photons from charged particles, and measure the charge of the latter up to $Z = 26$. The five faces of PSD envelope the five FIT sectors. Its design is being optimizing to achieve a high efficiency in the charged particles detection (> 99.98%), a high dynamic range to identify nuclei at least up to iron, a good segmentation to reduce the back-scattering particles from the calorimeter.

3.4 The silicon charge detector (SCD)

SCD is the HERD subdetector designed to accurately measure the cosmic ray absolute charge magnitude, thus separating chemical species from hydrogen to iron and beyond. SCD is composed of 4 x-y layers of single-sided silicon micro-strip sensors, for 8 independent ionization energy loss measurements per sector. It is the outermost detector to avoid early charge-change interactions in PSD and to reduce the systematic uncertainty on the reconstructed charge due to the nuclei fragmentation. It is highly segmented to minimize the unavoidable backscattered secondary particles coming from the calorimeter. The SCD charge measurement performance has been studied with a Monte Carlo simulation of the full detector. Several proton, helium, carbon, oxygen, silicon, and iron nuclei samples with energies from 10 GeV/nucleon to 1 TeV/nucleon and with isotropic incidence were simulated. It was found (Fig. 7) that the combined charge resolution of 8 layers is below 0.2 charge units for all samples. Design choices for SCD, like the strip pitch and the distance between SCD and PSD, will also be studied in order to minimize the effect of pile-up in charge measurement due to back-splash particles emerging from interactions in the calorimeter. Additionally, SCD is capable of 3D tracking and will be used together with FIT for the particle track reconstruction.

3.5 The transition radiation detector (TRD)

TRD [22] is needed to calibrate the response of the calorimeter to high energy hadronic showers. Space calorimeters can be calibrated on ground only up to 400 GeV using the CERN SPS beam. TRD is used to evaluate the CALO response from 1 TeV to 10 TeV. The TRD is conceived around a modular design, involving a transition radiation radiator and a thick gaseous electron multiplier (THGEM) with an effective area of 20×20 cm^2.

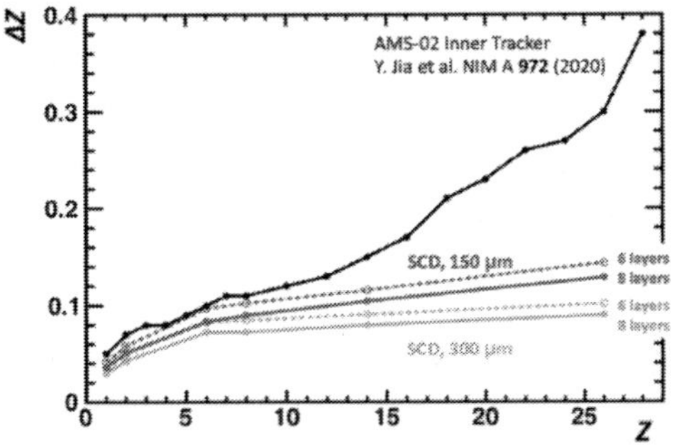

Figure 7: Simulated charge resolution of a SCD sector made of 6 and 8 layers of silicon micro-strip sensors with a thickness of 150 μm and 300 μm as a function of the nuclei charge *Z*. The charge resolution of the AMS inner tracker is also shown for comparison.

4 Conclusions

HERD will operate on board the China Space Station from 2027 for at least 10 years. The main scientific objectives of HERD include the indirect search for dark matter in the electron plus positron and gamma-ray spectra, the study of the gamma-ray sky, the detection of cosmic electrons and positrons, protons and heavier nuclei to understand the origin, acceleration and propagation of cosmic particles.

5 Acknowledgments

The author gratefully acknowledges the financial support from the Swiss National Science Foundation (SNSF-PZ00P2_193523).

References

[1] N. Aghanim, Y. Akrami, M. Ashdown, J. Aumont, C. Baccigalupi, M. Ballardini, A.J. Banday, R.B. Barreiro, N. Bartolo, S. Basak et al. (Planck Collaboration), Astronomy&Astrophysics **641**, A6 (2020)

[2] R.L. Workman, V.D. Burkert, V. Crede, E. Klempt, U. Thoma, L. Tiator, K. Agashe, G. Aielli, B.C. Allanach, C. Amsler et al., Progress of Theoretical and Experimental Physics **2022** (2022)

[3] B. Acharya, M. Actis, T. Aghajani, G. Agnetta, J. Aguilar, F. Aharonian, M. Ajello, A. Akhperjanian, M. Alcubierre, J. Aleksić et al. (CTA Collaboration), Astroparticle Physics **43**, 3 (2013)

[4] W.B. Atwood, A.A. Abdo, M. Ackermann, W. Althouse, B. Anderson, M. Axelsson, L. Baldini, J. Ballet, D.L. Band, G. Barbiellini et al. (Fermi-LAT Collaboration), The Astrophysical Journal **697**, 1071 (2009)

[5] L. Accardo, M. Aguilar, D. Aisa, B. Alpat, A. Alvino, G. Ambrosi, K. Andeen, L. Arruda, N. Attig, P. Azzarello et al. (AMS Collaboration), Physical Review Letters **113**, 121101 (2014)

[6] O. Adriani, G.C. Barbarino, G.A. Bazilevskaya, R. Bellotti, M. Boezio, E.A. Bogomolov, L. Bonechi, M. Bongi, V. Bonvicini, S. Bottai et al. (PAMELA Collaboration), Nature **458**, 607 (2009)

[7] G. Ambrosi, Q. An, R. Asfandiyarov, P. Azzarello, P. Bernardini, B. Bertucci, M.S. Cai, J. Chang, D.Y. Chen, H.F. Chen et al. (DAMPE Collaboration), Nature **552**, 63 (2017)

[8] K. Fang, X.J. Bi, P.F. Yin, The Astrophysical Journal **854**, 57 (2018)

[9] M. Aguilar, L. Ali Cavasonza, G. Ambrosi, L. Arruda, N. Attig, P. Azzarello, A. Bachlechner, F. Barao, A. Barrau, L. Barrin et al. (AMS Collaboration), Phys. Rev. Lett. **122**, 041102 (2019)

[10] M. Ajello, A. Allafort, M. Axelsson, L. Baldini, G. Barbiellini, M.G. Baring, D. Bastieri, R. Bellazzini, B. Berenji, E. Bissaldi et al. (Fermi-LAT Collaboration), The Astrophysical Journal **861**, 85 (2018)

[11] J. Aasi, B.P. Abbott, R. Abbott, T. Abbott, M.R. Abernathy, K. Ackley, C. Adams, T. Adams, P. Addesso, R.X. Adhikari et al. (LIGO Collaboration), Classical and Quantum Gravity **32**, 074001 (2015)

[12] F. Acernese, M. Agathos, K. Agatsuma, D. Aisa, N. Allemandou, A. Allocca, J. Amarni, P. Astone, G. Balestri, G. Ballardin et al. (Virgo Collaboration), Classical and Quantum Gravity **32**, 024001 (2015)

[13] H. Abe, T. Akutsu, M. Ando, A. Araya, N. Aritomi, H. Asada, Y. Aso, S. Bae, R. Bajpai, K. Cannon et al. (KAGRA Collaboration), Galaxies **10**, 63 (2022)

[14] M. Ahlers, K. Helbing, C.P. de los Heros, The European Physical Journal C **78**, 924 (2018)

[15] S. Adrián-Martínez, M. Ageron, F. Aharonian, S. Aiello, A. Albert, F. Ameli, E. Anassontzis, M. Andre, G. Androulakis, M. Anghinolfi et al. (KM3NeT Collaboration), Journal of Physics G: Nuclear and Particle Physics **43**, 084001 (2016)

[16] L. Fariña, L. Jouvin, J. Rico, N. Mori, F. Gargano, V. Formato, F. de Palma, C. Pizzolotto, J. Casaus, M.N. Mazziota et al., PoS **ICRC2021**, 651 (2021)

[17] L. Pacini, O. Adriani, Y. Bai, T. Bao, E. Berti, S. Bottai, W. Cao, J. Casaus, X. Cui, R. D'Alessandro et al., PoS **ICRC2021**, 066 (2021)

[18] O. Adriani, M. Antonelli, A. Basti, E. Berti, P. Betti, G. Bigongiari, L. Bonechi, M. Bongi, V. Bonvicini, S. Bottai et al., Journal of Instrumentation **17**, P09002 (2022)

[19] C. Perrina, P. Azzarello, F. Cadoux, Y. Favre, J.M. Frieden, D. La Marra, D. Sukhonos, X. Wu, PoS **ICRC2021**, 067 (2021)

[20] A. Sanmukh, S. Gómez, A. Comerma, J. Mauricio, R. Manera, A. Iraola, A. Sanuy, X. Wu, P. Azzarello, C. Perrina et al., in *2021 IEEE Nuclear Science Symposium and Medical Imaging Conference (NSS/MIC)* (2021), pp. 1–3

[21] D. Kyratzis, F. Alemanno, C. Altomare, P. Bernardini, P. Cattaneo, I. De Mitri, F. de Palma, L. Di Venere, M. Di Santo, P. Fusco et al., PoS **ICRC2021**, 054 (2021)

[22] B. Huang, H. Liu, X. Huang, M. Xu, Y. Dong, X. Wei, X. Liu, H. Feng, Q. Yang, J. Gu et al., Nuclear Instruments and Methods in Physics Research Section A: Accelerators, Spectrometers, Detectors and Associated Equipment **962**, 163723 (2020)

Ultra-high energy inverse Compton emission from Galactic electron accelerators

Mischa Breuhaus[1,*], *Joachim* Hahn[1,2], *Carlo* Romoli[1], *Brian* Reville[1], *Gwenael* Giacinti[1], *James Anthony* Hinton[1], and *Richard* Tuffs[1]

[1]Max-Planck-Institut für Kernphysik, Saupfercheckweg 1, 69117 Heidelberg, Germany
[2]CGI Darmstadt, Rheinstraße 95, 64295 Darmstadt, Germany

Abstract. It is generally held that >100 TeV emission from astrophysical objects unambiguously demonstrates the presence of PeV protons or nuclei, due to the unavoidable Klein–Nishina suppression of inverse Compton emission from electrons. However, in the presence of inverse Compton dominated cooling, hard high-energy electron spectra are possible. We show that the environmental requirements for such spectra can naturally be met in spiral arms, and in particular in regions of enhanced star formation activity, the natural locations for the most promising electron accelerators: powerful young pulsars. Leptonic scenarios are applied to gamma-ray sources recently detected by the High-Altitude Water Cherenkov Observatory (HAWC) and the Large High Altitude Air Shower Observatory (LHAASO). We show that these sources can indeed be explained by inverse Compton emission.

1 Introduction

Only recently we are starting to explore the γ-ray sky at ultra-high energies (UHE, >100 TeV). The first UHE γ-ray catalogue was obtained in 2020 by the High-Altitude Water Cherenkov Observatory (HAWC) collaboration [1] and contained only three sources. One year later the Large High Altitude Air Shower Observatory (LHAASO) collaboration detected 12 sources in this energy regime [2], providing spectra for three of them. The origin of these sources is of paramount interest since these energies probe the sources responsible for the Cosmic Ray (CR) 'knee' feature, which potentially marks the transition between Galactic and extragalactic CRs.

The observation of hard γ-ray spectra at these energies might be interpreted as being produced more likely by the collision of CRs with the ambient matter than by inverse Compton (IC) scattering by electrons because of the Klein-Nishina suppression. Yet, in radiation-dominated environments the situation is different, and hard IC spectra can still be possible [3–6]. Therefore, we investigated the possible leptonic origin of UHE γ-ray sources in detail and performed a leptonic modelling of three of the HAWC and LHAASO sources. This proceeding is mainly a summary of the results from our previous publications [6–8].

*e-mail: mischa.breuhaus@mpi-hd.mpg.de

2 Leptonic equilibrium spectra in radiation-dominated environments

The primary radiation mechanism for the production of UHE γ-rays by electrons or positrons is IC scattering. Unfortunately, unlike as for γ-ray production in hadronic collisions, the Klein-Nishina effects at high energies can prevent the occurrence of hard spectra. This is illustrated in the left panel of Figure 1, which shows the γ-ray spectrum produced by power-law distributed electrons IC scattering photons from a thermal radiation field with a temperature of $T = 50\,\mathrm{K}$. For the electrons, the spectrum is $\mathrm{d}N/\mathrm{d}E \propto E^{-\alpha}$, with $\alpha = 2$. In the Klein-Nishina regime, the spectral index of the γ-rays softens to $-(\alpha + 1)$. The only way to obtain a hard spectrum until higher energies is a hardening in the electron spectrum itself. Such a hardening occurs naturally in equilibrium situations in radiation energy dominated environments [3–5].

The right panel of Figure 1 shows an electron spectrum in equilibrium between injection and losses in a radiation energy dominated environment. The electron injection spectrum was an exponential cutoff power-law, $\mathrm{d}N/\mathrm{d}E \propto E^{-\alpha} \cdot \exp(-E/E_{\mathrm{cut}})$. The different cooling processes determine the shape. At energies below $\sim$100 TeV IC cooling is dominating. In the Klein-Nishina regime, the energy loss rate is lower and the electron spectrum hardens. This hardening can compensate for the softening of the resulting γ-ray emission. Eventually, synchrotron losses will dominate over IC energy losses in the Klein-Nishina regime, and the electron spectrum softens again. The transition energy E_{X} where IC and synchrotron losses are equal determines until which energy hard IC spectra can be produced. To be able to observe hard γ-ray spectra until 100 TeV values of $E_{\mathrm{X}} \gtrsim 100\,\mathrm{TeV}$ are required. This can only be achieved in environments where the parameter $\Xi_{\mathrm{IC}} \equiv U_{\mathrm{rad}}/U_{\mathrm{B}}$ is sufficiently high. Here, U_{rad} and U_{B} are the energy densities of the radiation field and the magnetic field.

The temperatures of the photon fields have an important influence on the required values of Ξ_{IC}. For lower temperatures, the Klein-Nishina transition occurs at higher energies. This implies that a smaller Ξ_{IC} value is sufficient to allow hard IC spectra until a certain energy. Hence, far infrared fields are the most important radiation fields which need to be taken into consideration. For typical Galactic dust temperatures between 10 and 50 K [9, 10] one needs values of $\Xi_{\mathrm{IC}} \sim 10$ and $\Xi_{\mathrm{IC}} \sim 100$ to achieve $E_{\mathrm{X}} = 100\,\mathrm{TeV}$. For the cosmic microwave background (CMB), already $\Xi_{\mathrm{IC}} \sim 3$ is sufficient. Since the CMB exists everywhere, when-

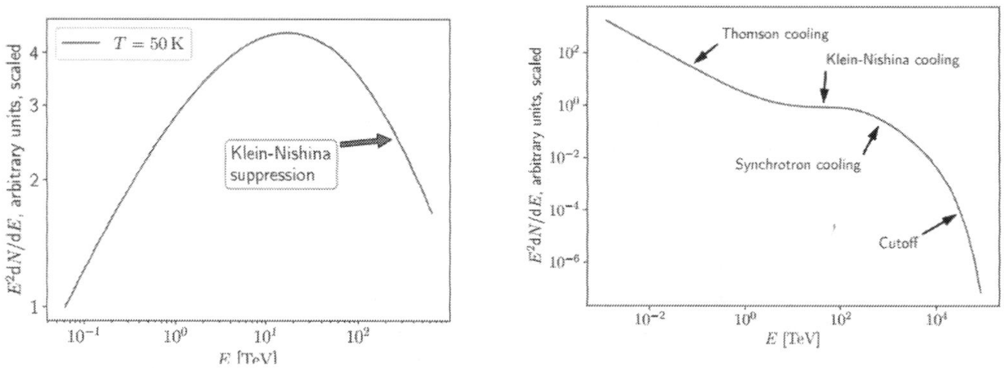

Figure 1. Left: Inverse Compton γ-ray spectrum from power-law distributed electrons with $\mathrm{d}N/\mathrm{d}E \propto E^{-2}$ with a radiation field of temperature $T = 50\,\mathrm{K}$. Right: Equilibrium electron spectrum in a radiation-dominated environment, for an injection spectrum $\mathrm{d}N/\mathrm{d}E \propto E^{-2}$.

ever the magnetic field drops below 1.8 μG the conditions to allow the production of hard IC spectra until ~100 TeV are fulfilled.

3 System constraints and potential Galactic environments

The acceleration of electrons to sufficiently high energies requires the magnetic fields to be sufficiently strong [e.g. 11]. This imposes a lower limit on the size of an accelerator for a given magnetic field strength. On the other hand, the strong radiation fields required to ensure dominance of radiation losses will also absorb the γ-rays, and the potential source could be opaque if it is too large. However, as we showed in [6], for magnetic fields below a few tens of μG a sufficiently large potential parameter space exists. The absorption by the interstellar Galactic radiation fields at 100 TeV is below 0.5 even in the most pessimistic case, but it can be more significant for higher γ-ray energies close to 1 PeV due to the CMB.

Possible PeV electron accelerators are pulsars with spin-down luminosities $\gtrsim 10^{36}$ erg s^{-1}. However, the strong magnetic fields will prevent values of $\Xi > 1$. Therefore, the pulsar termination shock has to be located far away from the central source where the magnetic fields are sufficiently low. Another possibility is that the electrons are accelerated in a high B-field region but diffuse sufficiently fast outwards without suffering significant energy losses to fill a larger volume where $\Xi > 1$. We demonstrated that this is indeed possible with reasonable diffusion coefficients, and therefore pulsars remain promising candidates for UHE IC sources.

Using the Galactic magnetic field model from [12, 13] and the radiation model from [14], we found that on large scales in the Milky Way the required environmental conditions can only be fulfilled at large Galactic radii or high above or below the disk where sources are scarce. However, the conditions can be very different locally. In star forming regions, far infrared energy densities ~100 eV cm^{-3} can be maintained over tens of pc [e.g. 15, 16]. Additionally, regions with low magnetic field values $B \lesssim 3$ μG such as expected in superbubbles [17] are ideal candidates too. Hence, many potential regions exist in the Galactic disk.

4 Leptonic modelling of HAWC and LHAASO sources

To investigate if the sources detected so far can be explained by leptonic scenarios, we first modelled the three sources detected by HAWC with simple equilibrium models. For the radiation field, we used the model from [14], allowing for an additional enhancement factor which we required to be compatible with data from the Infrared Astronomical Satellite (IRAS, [18]). Fixing the magnetic field to 3 μG and the power-law indices of the injection spectrum to $\alpha = 2$, reasonable models with enhancement factors below five can explain the data well above energies of 10 TeV where the equilibrium assumption is still valid [for details see 6]. All sources have a potential pulsar counterpart with sufficiently high spin-down power to account for the emission.

Additionally, we modelled the UHE LHAASO sources. Two of the three LHAASO sources for which the spectra are provided in [2] are counterparts to two of the HAWC sources. Hence, both datasets were modelled together. This time we took the finite source age into account and assumed that the electrons are only injected over the lifetime of the potential pulsar counterpart, which exists for all three sources. No enhancement of the large-scale Galactic radiation field was allowed. Free model parameters were fit to the data, while some of them such as the magnetic field or the power-law injection index were fixed in different fits. The data can be matched well, and reasonable resulting fit parameters were obtained in different scenarios [for details see 7]. With the γ-ray data from HAWC and LHAASO alone, the fit parameters can not be constrained very well, and additional information about the source environments is needed for further conclusions. However, each source can be easily explained with leptonic scenarios, and a leptonic origin of at least some of the sources is likely.

5 Summary and conclusion

We showed that environments required to produce hard IC spectra could exist in many local regions in the Milky Way where the far infrared radiation energy density is enhanced and/or the B-fields are low. Hard spectra until 100 TeV are possible if IC energy losses dominate until sufficiently high energies. High spin-down power pulsars coincident with star-forming regions and superbubbles are ideal candidates. Superbubbles, star formation, and the creation of new neutron stars are expected to occur together, and it is likely that high spin-down pulsars exist in suitable regions.

The HAWC and LHAASO sources investigated here can be modelled with reasonable leptonic scenarios. However, there is a redundancy in the model parameters, even for the simple models used here. Multiwavelength data are crucial to be able to distinguish leptonic from hadronic sources, even at UHE. From the 12 sources detected by LHAASO, only two do not have a potential pulsar counterpart. It might therefore be likely, that several of these sources are indeed powered by pulsars and have a leptonic origin. IC sources are likely an important contributor to the γ-ray sky at energies above 100 TeV.

References

[1] A.U. Abeysekara, A. Albert, R. Alfaro, J.R. Angeles Camacho et al. (HAWC Collaboration), Phys. Rev. Lett. **124**, 021102 (2020)
[2] Z. Cao, F. Aharonian, Q. An et al., Nature**594**, 33 (2021)
[3] G.R. Blumenthal, R.J. Gould, Reviews of Modern Physics **42**, 237 (1970)
[4] F.A. Agaronyan, A.S. Ambartsumyan, Astrophysics **23**, 650 (1985)
[5] A.A. Zdziarski, J.H. Krolik, ApJ**409**, L33 (1993)
[6] M. Breuhaus, J. Hahn, C. Romoli, B. Reville, G. Giacinti, R. Tuffs, J.A. Hinton, ApJ**908**, L49 (2021), 2010.13960
[7] M. Breuhaus, B. Reville, J.A. Hinton, A&A**660**, A8 (2022), 2109.05296
[8] M. Breuhaus, J. Hahn, C. Romoli, B. Reville, G. Giacinti, R. Tuffs, J. Hinton, *Ultra-high Energy Inverse Compton Emission from Galactic Electron Accelerators*, in *Proceedings of 37th International Cosmic Ray Conference — PoS(ICRC2021)* (2021), Vol. 395, p. 932
[9] J.P. Bernard, D. Paradis, D.J. Marshall, L. Montier et al., A&A**518**, L88 (2010)
[10] J. Zhu, M. Huang, A&A**564**, A111 (2014)
[11] A.M. Hillas, ARA&A**22**, 425 (1984)
[12] R. Jansson, G.R. Farrar, ApJ**757**, 14 (2012)
[13] R. Jansson, G.R. Farrar, ApJ**761**, L11 (2012)
[14] C.C. Popescu, R. Yang, R.J. Tuffs, G. Natale, M. Rushton, F. Aharonian, MNRAS**470**, 2539 (2017)
[15] M.A. Dopita, B.A. Groves, J. Fischera, R.S. Sutherland, R.J. Tuffs, C.C. Popescu, L.J. Kewley, M. Reuland, C. Leitherer, ApJ**619**, 755 (2005)
[16] C.C. Popescu, R.J. Tuffs, M.A. Dopita, J. Fischera, N.D. Kylafis, B.F. Madore, A&A**527**, A109 (2011), 1011.2942
[17] M.J. Korpi, A. Brandenburg, A. Shukurov, I. Tuominen, A&A**350**, 230 (1999)
[18] G. Neugebauer, H.J. Habing, R. van Duinen, H.H. Aumann et al., ApJ**278**, L1 (1984)

EPJ Web of Conferences **280**, 03001 (2023) https://doi.org/10.1051/epjconf/202328003001
RICAP-22

KM3NeT/ARCA Sensitivity to Starburst Galaxies

A. Ambrosone[1,2], *W.I.* Ibnsalih[2,3,*], A. Marinelli[1,2,5], G. Miele[1,2,4], P. Migliozzi[2], M. R. Musone[2,3]On behalf of the KM3NeT Collaboration

[1]Dipartimento di Fisica "Ettore Pancini", Università degli studi di Napoli "Federico II", Complesso Univ. Monte S. Angelo, I-80126 Napoli, Italy
[2]INFN - Sezione di Napoli, Complesso Univ. Monte S. Angelo, I-80126 Napoli, Italy
[3]Università degli studi della Campania Luigi Vanvitelli
[4]Scuola Superiore Meridionale, Università degli studi di Napoli "Federico II", Largo San Marcellino 10, 80138 Napoli, Italy
[5]INAF-Osservatorio Astronomico di Capodimonte, Salita Moiariello 16, I-80131 Naples, Italy

Abstract. In this work, the expectations of the full detector KM3NeT/ARCA are presented for particular starburst galaxies signals. For the point-like search approach, we considered the most promising local starburst galaxies to be observed as point-like neutrino excesses: NGC 1068, the Small Magellanic Cloud and the Circinus Galaxy. In both cases, we provide the energy-integrated sensitivity for two ARCA building blocks in the energy range 100 GeV – 10 PeV. In the diffuse scenario, both track and shower events were considered. For the point like analysis, only ν_μ charge current events were taken into account. Interestingly, ARCA has the potential to constrain the selected phenomenological scenarios, providing evidence of the link between star-forming processes and hadronic emissions.

1 Introduction

The KM3NeT/ARCA telescope is an under water Cherenkov telescope under construction in the Mediterranean sea [1]. The telescope is composed of two different detectors: ORCA and ARCA. Being sensitivity to low energy neutrinos, ORCA is devoted to neutrino physics studies. ARCA is devoted to high-energy neutrino astronomy, being sensitive to neutrinos above $\sim 10^2$ GeV. In this contribution the integrated sensitivity of ARCA to constrain high-energy neutrinos from Starburst Galaxies is obtained, using a cut count approach. Both a diffuse and a point-like analysis is presented. From the point-like side, the most promising point-like sources were considered. In particular: NGC 1068, since IceCube has found strong evidence for its neutrino emissions [2], and also the Small Magellanic Cloud (SMC) and Circinus since [3] has found that ARCA could possibly see evidence of the star forming activity occuring in their cores. For the diffuse study, the spectra of [4] were considered. This model implemented a blending of spectral indices to exploit the variability of this parameter along the source class. This paper is organised as follows: in Sec. 2, the sensitivity calculation and the diffuse analysis are discussed; in Sec. 3 the sensitivities for the point-like search are presented. Finally, in Sec. 4 the discussion and the conclusions are reported.

*e-mail: walid.idrissiibnsalih@unicampania.it

2 Sensitivity Calculation and Diffuse Analysis

The Feldman and Cousins method was used to evaluate of the 90% C.L. upper limits, for further details see [5]. The sensitivity is defined as $\Phi_{90} = (\bar{\mu}_{90}/n_s)\Phi_\nu$, where $\bar{\mu}_{90}$ is the average upper limit, n_s the signal events predicted by the signal neutrino flux Φ_ν. $\bar{\mu}_{90}/n_s$ is defined as the Model Rejection Factor (MRF). For the diffuse analysis, the energy range [100 GeV – 10 PeV] was considered. For the background, atmospheric muons were simulated with MUPAGE ([6]) and gSeaGen([7]) was used for the atmospheric neutrinos. The spectrum provided by [4] was used for the signal which represents the expected neutrino diffuse signal of SBGs. In order to reject background and minimize the sensitivity, dedicated track-like and cascade-like event selections were performed. For the **track-like** events, only upgoing events were selected: $\theta < 100°$, where θ is the Zenith angle of the reconstruction event. To reject poorly reconstructed tracks, additional criteria were applied: $\beta_0 < 1$ deg, where β_0 is the error on the reconstructed angle; Len > 300 m, where Len is the reconstructed track-length; Len/$n_{\rm hits}$ > 0.7, where n_{hits} is the number of hits. In order to select the good track up-going events and reject atmospheric muons, a boosted decision tree (BDT) was implemented. The BDT training was based in the main reconstructed variables (e.g. Len, β_0, n_{hits}), using the KM3NeT/ARCA115 Monte Carlo simulated events (atmospheric muons and neutrinos). For the final event selection, BDT score conditions were used: for $80° < \theta < 100°$, $BDT_{score} > 0.9$ - for $\theta < 80$ deg, $BDT_{score} > 0.4$. For the **cascade-like** event selection, several conditions were applied: the events containment required to be inside the effective volume of the detector ($R_{det} < 600, Z_{det<650}$) , short events length (Len<300 meters) and the selection on quality reconstruction variables ($n_{\rm hits} > 450, L < -500$). In addition, A BDT trained and optimized for cascade events (BDT_{casc}) was applied. Regarding the direction of the events, a selection of BDT scores was applied:

- $cos(\theta) > 0.6$: $BDT_{casc} > -1.1$
- $0.2 < cos(\theta) < 0.6$: $BDT_{casc} > -0.9$
- $-0.2 < cos(\theta) < 0.2$: $BDT_{casc} > -0.8$
- $-0.6 < cos(\theta) < -0.2$: $BDT_{casc} > 0.8$
- $cos(\theta) < -0.6$: $BDT_{casc} > 1.1$

Fig. 1 summarizes the results. The computed sensitivity is compared with two different SBG scenarios: on the left the expected signal was calculated considering HESE (IceCube) and Fermi-LAT EGB, and on the right using CASCADE (IceCube) and Fermi- LAT EGB (see [4] for furher details). The sensitivity is shown in the range where 90% of the signal is concentrated. The comparison with the theoretical expectations shows that ARCA is sensitive for such a spectrum and in few years of data taking it will put several constrain on such a scenario.

3 Point-Like Analysis

For the point-like analysis, only ν_μ and $\bar{\nu}_\mu$ in charged current interactions (track-like events) were considered. The Monte Carlo simulations events were the same as for the diffuse analysis for the background events. The dedicated simulation events for the signal were generated using the gSeaGen code [7]. Circinus and NGC 1068 were simulated as point-like sources, while SMC was considered as a 0.5° extended source, compatible with the emissions coming from its nucleus [3]. In order to evaluate the sensitivity, a binned search was employed to evaluate the number of events through the region of interest (RoI). Therefore, α (the angular distance between the position of the source and the coordinates of the reconstructed track) is used to minimize the MRF. For the signal, only events in the RoI were taken into account. For the background, the whole corresponding declination band is considered and then rescaled

https://doi.org/10.1051/epjconf/202328003001

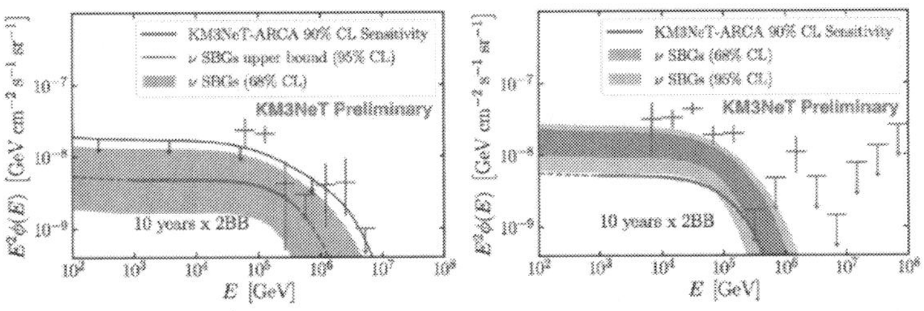

Figure 1. Left The 90% C.L. sensitivity (red curve) is compared with the corresponding theoretical 1σ uncertainty band of the SBG spectral energy distribution according to the analysis of [4]. This band has been obtained by fitting the Fermi-LAT EGB and the IceCube HESE data. the IceCube HESE measurements [8] are also shown for reference. **Right** The sensitivity is compared with the corresponding theoretical $1-2\sigma$ uncertainty band of the SBG spectral energy distribution according to the analysis of [4], considering EGB Fermi-lat and (6-year) of IceCube CASCADE samples. The IceCube cascades measurements ([9]) are shown for reference.

to be consistent with the width of the RoI. The rescaling factor is $\alpha/4 = \Omega_{\mathrm{RoI}}/\Omega_{db}$, where $\Omega_{\mathrm{RoI}} \simeq \pi\alpha^2$ and $\Omega_{\mathrm{db}} \simeq 2\pi 2\alpha$. Finally, the MRF is minimized as a function of several parameters in order to obtain the optimized value for the sensitivity. Figs. 2-4 report the results for the three sources analyzed.

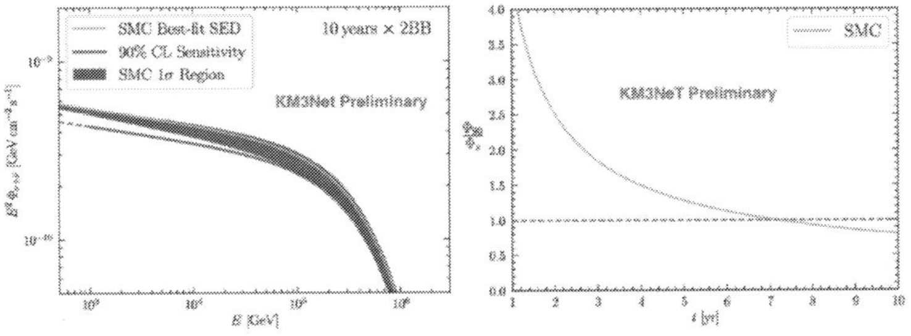

Figure 2. Left sensitivity compared with the theoretical 1 σ uncertainty band of the source (according to the analysis of [3]). **Right** Sensitivity as a function of time. The detector will be able to put strict constraints after 7 years of data taking.

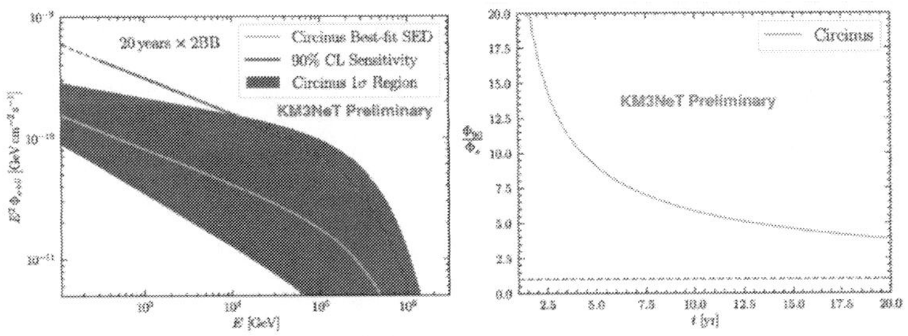

Figure 3. Left The integrated sensitivity compared with the theoretical 1 σ uncertainty band of the source (according to the analysis of [3]). The plot refers to 20 years of data taking for 2 Building Blocks. **Right** The ARCA sensitivity as a function of time. The detector is not able to discriminate the SBG scenario in the best-fit scenario. It can trace the upper limit of the expected emission.

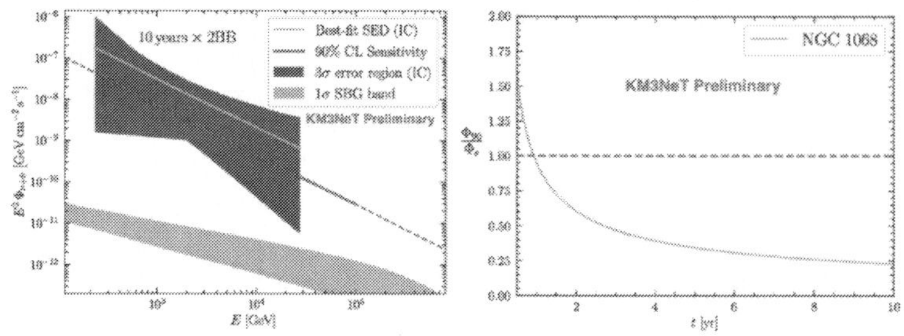

Figure 4. Left The sensitivity compared with the theoretical 1σ uncertainty band of the source (according to the analysis of [3]) as well as IceCube error region [2]. **Right** The ARCA sensitivity as a function of time. The detector will be able in 1 year to constrain the scenario inferred by IceCube.

In particular, the left panels of figures 2-4 show the sensitivity compared with the spectrum after 10 years (20) of data taking of 2 ARCA building blocks for SMC and NGC 1068 (Circinus). The right panel shows the obtained MRF scaling with time in the best-fit scenario. In each case, the results show that in few-years ARCA will be able to test these hadronic scenario posing strict constraints on the hadronic budget of Starburst Galaxies.

4 Conclusions

All the results point to the conclusion that KM3NeT/ARCA will be fundamental not only to unveil the origin of the astrophysical neutrino spectra, but also, it is going to test if the star forming activity can be traced by neutrino production. This will be an invaluable step forward for neutrino astronomy.

References

[1] S. Adrian-Martinez et al. Letter of intent for KM3NeT 2.0. J. Phys. G, 43(8):084001, 2016.

[2] M. G. Aartsen et al. Time-Integrated Neutrino Source Searches with 10 Years of IceCube Data. Phys. Rev. Lett., 124(5):051103, 2020

[3] Antonio Ambrosone et al. Could Nearby Star-forming Galaxies Light Up the Pointlike Neutrino Sky? Astrophys. J. Lett., 919(2):L32, 2021.

[4] Antonio Ambrosone, et al. Starburst galaxies strike back: a multi-messenger analysis with Fermi-LAT and IceCube data. Mon. Not. Roy. Astron. Soc., 503(3):4032–4049, 2021.

[5] Gary J. Feldman and Robert D. Cousins. A Unified approach to the classical statistical analysis of small signals. Phys. Rev. D, 57:3873–3889, 1998.

[6] G. Carminati et al.. Atmospheric MUons from PArametric formulas: A Fast GEnerator for neutrino telescopes (MUPAGE). Comput. Phys. Commun., 179:915–923, 2008.

[7] S. Aiello et al. gSeaGen: The KM3net GENIE-based code for neutrino telescopes. Computer Physics Communications, 256:107477, nov 2020.

[8] Austin Schneider. Characterization of the Astrophysical Diffuse Neutrino Flux with IceCube High- Energy Starting Events. PoS, ICRC2019:1004, 2020.

[9] M. G. Aartsen et al. Search for astrophysical sources of neutrinos using cascade events in IceCube. Astrophys. J., 846(2):136, 2017.

Gamma-Ray and Neutrino Emissions from Starforming and Starburst Galaxies

Antonio Ambrosone[1,2,*], *Marco* Chianese[1,2], *Damiano F.G.* Fiorillo[5], *Antonio* Marinelli[1,2,4], and *Gennaro* Miele[1,2,3]

[1]Dipartimento di Fisica "Ettore Pancini", Università degli studi di Napoli "Federico II", Complesso Univ. Monte S. Angelo, I-80126 Napoli, Italy
[2]INFN - Sezione di Napoli, Complesso Univ. Monte S. Angelo, I-80126 Napoli, Italy
[3]Scuola Superiore Meridionale, Università degli studi di Napoli "Federico II", Largo San Marcellino 10, 80138 Napoli, Italy
[4]INAF-Osservatorio Astronomico di Capodimonte, Salita Moiariello 16, I-80131 Naples, Italy
[5]Niels Bohr International Academy, Niels Bohr Institute, University of Copenhagen, Copenhagen, Denmark

Abstract. Experimental observations have demonstrated a strong correlation between the star formation rate and the gamma-ray lumosities of starforming and starburst galaxies (SFGs and SBGs). However, the real origin of these emissions is still under debate. In this contribution, we present several updates on their non-thermal radiations, revisiting both their point-like and cumulative (diffuse) emission properties. From the point-like side, we discuss the potentialities of future neutrino (KM3NeT/ARCA, IceCube-gen2) telescopes to quantitively scrutinize their expected properties from different cosmic-ray transport models. From the diffuse perspective, we investigate a model based on a data-driven blending of spectral indexes, hence taking into account the changes in the properties of individual emitters. Strikingly, SFGs and SBGs can explain 25% (up to 40%) of the diffuse High-Energy Starting Events (HESE) data, without overshooting the gamma-ray limits regarding non-blazar sources.

1 Introduction

Starforming and Starburst Galaxies are astrophysical sources charactherised by a high star formation rate (see [1–5]). This means that many stars are born and die in their astrophysical environment. This activity is principally expected to occur in their core, which is also called starburst nucleus (SBN) [1, 2]. Consequently, supernova explosions (SNe) are expected to inject a great amount of high-energy cosmic rays (protons with energies up to ~ 1 PeV) (see [3–5] and Refs. therein for other details). Furthermore, a gas density can be as high as 10^{2-3} cm^{-3} inside SBNi (see [5] for more details). Therefore, SFGs and SBGs should emit non-thermal particles such as gamma-rays and neutrinos through hadronic collisions between high-energy cosmic rays partcles and molecular gas particles. In this contribution, we describe how current and future experiment can test the hadronic production of these sources. In particular, we follow the model put forward by [1], which considers cosmic ray transport dominated by advection phenomena. This is phenomenologically justifed by the fact that gamma-ray spectra

*e-mail: antonio.ambrosone@unina.it

of most SFGs and SBGs are hard power-laws (see [6] for details about the observed spectra) which, to some extent, implies that the escape timescale is energy-independent. While, theoretically, it is justified because of the expected high-degree magnetic turbulence [7] which leads to a minor role for diffusion. From the point-like side, we analyze photons data from [6], namely 10-yr *Fermi*-LAT spectral energy distribution data. We also use VERITAS [8] and HESS data [9] respectively for M82 and NGC 253. For Circinus Galaxy, we use data from [10]. We analyze these data with the intention of evaluating the neutrino contribution of these sources and test if neutrino telescopes can trace these emissions. From the diffuse analysis side, we use the calorimetry assumption (see [2, 3] for details) to calculate the amount of gamma-rays and neutrinos coming to the Earth from all these sources in the Universe. The proceeding is organized as follows: in Sec. 2, we discuss the cosmic-ray transport mechanisms as well as the point-like analysis; in sec. 3, we discuss the diffuse contribution of SFGs and SBGs. Finally, in sec. 4, we draw our conclusions.

2 Cosmic-Ray Transport and Point-Like Analysis

The emissions of SBGs are usually regarded as steady because of the starburst timescale duration. In this context, it is possible to use a simple leaky-box model to describe the cosmic ray transport inside the nucleus. Therefore (see [1–5]):

$$\frac{f(p)}{T_{\rm loss}} + \frac{f(p)}{T_{\rm diff}} + \frac{f(p)}{T_{\rm adv}} = Q(p) \tag{1}$$

where $f(p)$ is the CR momentum distribution inside the nuclei, $Q(p)$ is the injection rate from SNe. $T_{\rm loss}$, $T_{\rm diff}$ and $T_{\rm adv}$ are the energy loss, diffusion and advection timescales, respectively. $Q(p)$ follows a power-law spectrum with α as a spectral index and an exponential cut-off. The neutrino and gamma-ray are mainly produced by the pion decays and we parametrize these emissions using the delta-function approximation, assuming pions carry 17% of the kinetic energy of the parent high-energy proton. We analize every known (nearby) SFG, using *Fermi*-LAT data. We fit the gamma-ray data with this model in order to calculate the structural parameters needed to calculate the neutrino fluxes and test the potentiality of current and upcoming neutrino telescopes to constrain such scenarios. Fig. 1 summarizes our main result (see [4] for more details). It compares the 1 TeV neutrino SBG flux from our estimate with the IceCube [11] and the expected sensitivity of KM3NeT/ARCA detector [12] and ICeCube Gen2. Importantly, SMC and Circinus cores emissions could be constrained by 6 years of data taking of the upcoming KM3NeT detector [13]. Furthermore, the neutrino excess detected from the direction of NGC 1068 cannot be generated by SBG activity. We stress that we find, along with [6], that gamma-ray emissions of NGC 1068 to be consistent with its SBG activity within the calorimetric limit. However, this cannot exclude further contamination to these data from hot-corona and jet activities.

3 Diffuse Analysis

In this section, we analyze the expected diffuse gamma-ray and neutrino contributions of these sources. In particular, we exploit the fact that the higher the star formation rate, the more calorimetric these source are going to be [2, 3]. Indeed, above a critical star formation rate value ($\simeq 1\,{\rm M_\odot\,yr^{-1}}$), SFGs are expected to be calorimetric [2]. In this scenario, it is possible to set a prototype galaxy (in this case M82) and then the generic flux of a source can be considered as $f_{SBG}(E,\alpha,\psi) = \frac{\psi}{\psi_{M82}} f_{M82}(E,\alpha)$ [2, 5]. Therefore, the diffuse emissions will predominatly depend on the distribution of the spectral shape as well as on the maximal energy reachable by protons in the SBN environment. We assume a maximal energy of 10 PeV.

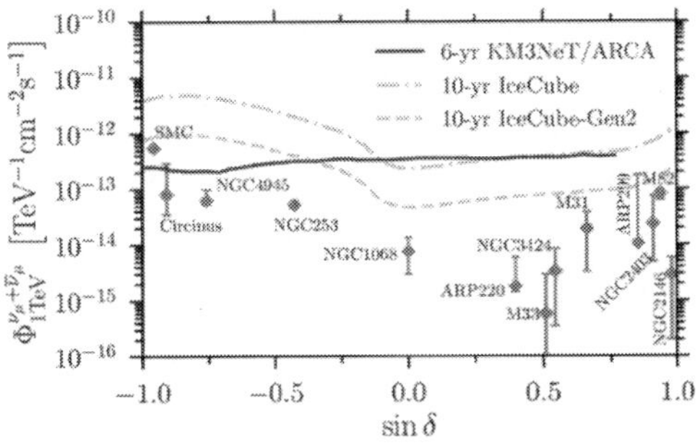

Figure 1. 1 TeV neutrino spectra normalizations compared with KM3NeT, IceCube and IceCube gen 2 sensitivities. Figure taken from [4].

For the α distribution, we consider the gaussian distribution considering the 13 best-fit values coming from point-like fit we have performed in the point-like analysis. Fig. 2 summarises the result comparing the Isotropic Gamma-ray background [14] and IceCube data [15] with the corresponding expected SBG spectra (see [5] for more details). Interestingly, we find

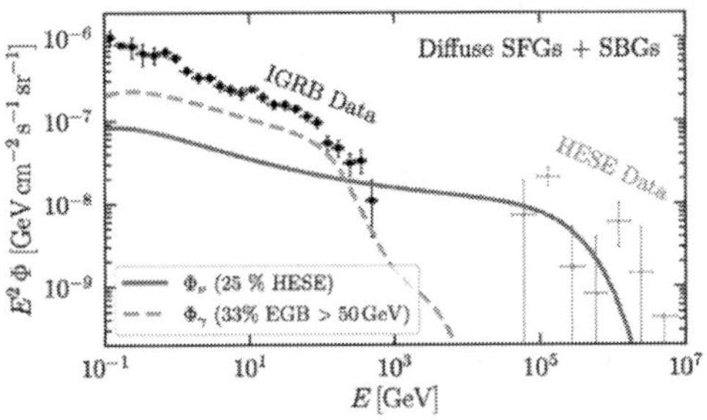

Figure 2. Diffuse Gamma-ray and Neutrino SBG spectra compared with IRGB [14] and HESE [15] data. Figure taken from [5].

that SFGs and SBGs give an important contribution to the Isotropic Gamma-ray Background light (IGRB), while being consistent with the Extragalactic Gamma-ray Background (EGB) constraints on non-blazar sources above 50 GeV ($\leq 30-40\%$) (see [16]). Moreover, the neutrino spectrum is expected to power about 25% of the HESE data. Indeed, the spectral index blending provides a hardening on the spectrum leading to a greater neutrino contribution. In Ref. [3], we also provided a multi-component analysis, using other sources (along with SFGs and SBGs) contributions such as blazars and radio galaxies. We have demonstrated that using a blending scenario, SFGs and SBGs can contribute to 40% of IceCube data at 2σ level.

4 Conclusions

In this proceeding, we have reviewed the gamma-ray and neutrino emission properties of SFGs and SBGs. In particular, using the state-of-the-art cosmic ray transport mechanisms scenario, we have provided a forecast for upcoming neutrino telescopes to understand if they can trace star formation activity. We obtained that the cores of SMC and Circinus could be constrained by future neutrino observations. This will be an invaluable test of such hadronic scenario. From the diffuse point of view, we have used the calometric scenario to evaluate the diffuse gamma-ray and neutrino spectra and compared with the corresponding IGRB and HESE data. We have obtained that SFGs and SBGs can explain a considerable fraction of IGRB data and also a part of IceCube data.

This work was partially supported by the research grant number 2017W4HA7S "NAT-NET: Neutrino and Astroparticle Theory Network" under the program PRIN 2017 funded by the Italian Ministero dell'Università e della Ricerca (MUR). The authors also acknowledge the support by the research project TAsP (Theoretical Astroparticle Physics) funded by the Istituto Nazionale di Fisica Nucleare (INFN).

References

[1] E. Peretti, P. Blasi, F. Aharonian and G. Morlino, Mon. Not. Roy. Astron. Soc. **487** (2019) no.1, 168-180 doi:10.1093/mnras/stz1161 [arXiv:1812.01996 [astro-ph.HE]].

[2] E. Peretti, P. Blasi, F. Aharonian, G. Morlino and P. Cristofari, Mon. Not. Roy. Astron. Soc. **493** (2020) no.4, 5880-5891 doi:10.1093/mnras/staa698 [arXiv:1911.06163 [astro-ph.HE]].

[3] A. Ambrosone, M. Chianese, D. F. G. Fiorillo, A. Marinelli, G. Miele and O. Pisanti, Mon. Not. Roy. Astron. Soc. **503** (2021) no.3, 4032-4049 doi:10.1093/mnras/stab659 [arXiv:2011.02483 [astro-ph.HE]].

[4] A. Ambrosone, M. Chianese, D. F. G. Fiorillo, A. Marinelli and G. Miele, Astrophys. J. Lett. **919** (2021) no.2, L32 doi:10.3847/2041-8213/ac25ff [arXiv:2106.13248 [astro-ph.HE]].

[5] A. Ambrosone, M. Chianese, D. F. G. Fiorillo, A. Marinelli and G. Miele, Mon. Not. Roy. Astron. Soc. **515** (2022) no.4, 5389-5399 doi:10.1093/mnras/stac2133 [arXiv:2203.03642 [astro-ph.HE]].

[6] M. Ajello, M. Di Mauro, V. S. Paliya and S. Garrappa, Astrophys. J. **894** (2020) no.2, 88 doi:10.3847/1538-4357/ab86a6 [arXiv:2003.05493 [astro-ph.GA]].

[7] T. A. Thompson, E. Quataert, E. Waxman, N. Murray and C. L. Martin, Astrophys. J. **645** (2006), 186-198 doi:10.1086/504035 [arXiv:astro-ph/0601626 [astro-ph]].

[8] VERITAS Collaboration and 92 colleagues 2009. A connection between star formation activity and cosmic rays in the starburst galaxy M82. Nature 462, 770–772. doi:10.1038/nature08557

[9] H. Abdalla *et al.* [H.E.S.S.], Astron. Astrophys. **617** (2018), A73 doi:10.1051/0004-6361/201833202 [arXiv:1806.03866 [astro-ph.HE]].

[10] M. Hayashida, Ł. Stawarz, C. C. Cheung, K. Bechtol, G. M. Madejski, M. Ajello, F. Massaro, I. V. Moskalenko, A. Strong and L. Tibaldo, Astrophys. J. **779** (2013), 131 doi:10.1088/0004-637X/779/2/131 [arXiv:1310.1913 [astro-ph.HE]].

[11] M. G. Aartsen *et al.* [IceCube], Phys. Rev. Lett. **124**, no.5, 051103 (2020) doi:10.1103/PhysRevLett.124.051103 [arXiv:1910.08488 [astro-ph.HE]].

[12] S. Aiello *et al.* [KM3NeT], Astropart. Phys. **111**, 100-110 (2019) doi:10.1016/j.astropartphys.2019.04.002 [arXiv:1810.08499 [astro-ph.HE]].

[13] S. Adrian-Martinez *et al.* [KM3Net], J. Phys. G **43**, no.8, 084001 (2016) doi:10.1088/0954-3899/43/8/084001 [arXiv:1601.07459 [astro-ph.IM]].

[14] M. Ackermann *et al.* [Fermi-LAT], Astrophys. J. **799** (2015), 86 doi:10.1088/0004-637X/799/1/86 [arXiv:1410.3696 [astro-ph.HE]].

[15] R. Abbasi *et al.* [IceCube], Phys. Rev. D **104** (2021), 022002 doi:10.1103/PhysRevD.104.022002 [arXiv:2011.03545 [astro-ph.HE]].

[16] M. Lisanti, S. Mishra-Sharma, L. Necib and B. R. Safdi, Astrophys. J. **832** (2016) no.2, 117 doi:10.3847/0004-637X/832/2/117 [arXiv:1606.04101 [astro-ph.HE]].

The Einstein Telescope: status of the project

Antonino Chiummo[1,*] for the ET Collaboration

[1]European Gravitational Observatory - EGO , Cascina (PI) - Italy

Abstract. After the groundbreaking first detection of Gravitational Waves by the LIGO/Virgo Collaborations, the design of the third generation of gravitation wave interferometric antennas has gained a substantial interest and momentum. The Cosmic Explorer in the US and the European Einstein Telescope (ET) are the instruments which will take the legacy of the current generation and further advance the reach of the GW antennas, both in terms of explored Universe and of type of accessible sources. The talk focused on the status of the design of the Einstein Telescope, its technical challenges and the expected implications to the Gravitational Wave Astronomy progress.
The paper reflects the content and the status of the ET project at the time of the talk given at RICAP22.

1 Introduction

A new Era in the observation of the Universe was opened by the first historical detection of a Gravitational Wave (GW) signal by second generation (2G) interferometric antennas [1]. Furthermore, the dawn of the Multimessenger Astronomy broke when the GW signal from a binary Neutron Star (BNS) cohalescence was detected, with the subsequent observation of an electromagnetic counterpart [2] in the region of sky pinpointed by the LIGO-Virgo detectors. Many other detections took place in the first three observation runs of the 2G detectors, with a large impact in both Astrophysics and Fundamental Physics, and more science advance is expected for the next observation runs of the current generation.
Nevertheless, such a huge success has to be considered just the first step in the observation of the Universe with GWs. Indeed, third-generation (3G) GW detectors, such as the Einstein Telescope (ET) [3] are already following the path from conceptual design to realization, and will allow the full realization of the potential of the GW Astronomy revolution, thanks to an order of magnitude better sensitivity and a wider accessible frequency band with respect to 2G detectors.
The interest for this next generation detectors has gained of course interest and momentum following the successful outcome of the 2G antennas, but the visionary path for their realization started long before the historical first GW detection [4], a path which brought to the current funding of the ET project and the creation of the ET collaboration at 2022 summer ET symposium, now counting more than 1200 members in 80 research units spread across Europe and Asia.

*e-mail: antonino.chiummo@ego-gw.it

2 The Einstein Telescope in a nutshell

2.1 The Observatory

The Einstein Telescope is meant to be not only a 3G detector but a real GW observatory. It will consist in a new underground (-100 m) infrastructure capable to host future upgrades for decades without limiting the observation capabilities. The sensitivity of the detector is aimed to be at least 10 times better than the nominal 2G detectors on a large fraction of the detection frequency band. Moreover, the design will implement a dramatic improvement in sensitivity in the low frequency range (1 Hz to 10 Hz). The infrastructure shall provide high reliability and improved observation capability, including GW polarization disentanglement.

2.2 ET science case

The ET science will span Astrophysics, Fundamental Physics and Cosmology as well. Accessible topics where ET will contribute include the study of Black Hole properties, their origin (stellar vs. primordial), evolution, demography. The same for the Neutron Star properties: it will be possible to study their interior structure (QCD at ultra-high densities, exotic states of matter), and demography. Moreover, the Einstein Telescope will greatly enhance the possibility of Multi-band and multi-messenger Astronomy, for instance by increasing joint GW/EM observations (e.g. GRB, kilonova, etc.). Detections of GW emission coming new astrophysical sources, such as core-collapse supernovae or isolated neutron stars will also become much more likely. The contributions of ET will span also fundamental Physics and Cosmology: it will allow investigations on dark energy equation of state, modified GW propagation, Stochastic backgrounds of cosmological origin and much more [5].

Some are just an evolution of what can be done with the 2G detectors, but ET sensitivity will provide insight in very extreme events with a high signal-to-noise ratio (SNR), and improve the multimessenger ability of the current detectors. In particular, the contribution of ET will be twofold: it will be a discovery machine on one side, because ET will explore almost the entire Universe listening the GWs emitted by black holes back to the dark ages after the Big Bang; and a precision measurement observatory on the other side, since ET will detect, with high SNR, hundreds of thousands of coalescences of binary systems of Neutron Stars per year, revealing the most intimate structure of the nuclear matter in their nuclei.

2.3 ET design and technological challenges

Unsurprisingly, in order to realize these challenging science goals, very tight requirements and design specifications are needed, which translates in tough technological challenges. ET is requested to operate in a wide frequency range (1 Hz to 10 Hz), to have an unprecedented sensitivity for massive black holes (focus on the low frequency part of the detection bandwidth), localisation capability, (more) Uniform sky coverage, polarisation disentanglement, high duty cycle, and high SNR. These requirements suggested a multi-interferometer design to extend the detection bandwidth (Xylophone), to go underground to provide a more quiet environment, to use cryogenic temperature for the test masses, to employ multi-detectors with different orientations for sky coverage and polarization disentanglement, and to the obvious choice of longer arms (10 Km) to increase the strain sensitivity. The Xylophone approach means that for each of the detectors in the multi-detector scheme, two different interferometers are realized: one more sensitive to the low-frequency band (LF), and the other to the high frequency (HF) region of the detection bandwidth. The requirements for these two interferometers are different and so their technological challenges. For the LF interferometer the thermal noise of coating and suspensions will play a major role, so a cryogenic temperature for the test masses is foreseen. This implies changing the substrate material of the test masses from Fused Silica to Silicon (or Sapphire) and change the wavelength of the laser, with the related new technologies to be developed. Furthermore, new seismic isolation has to be developed to cope with more stringent requirements at low frequency. Instead, the HF interferometer will require very high laser power, and this in turn will pose severe requirements to the ability to control thermal aberrations, just to cite some of the problems to face. A complete and updated technical design can be found in [6].

3 A site for ET

The site where ET will be built will play a non-negligible role for the ultimate performances of the detector in terms of noise. After a wider survey, basically two sites are still valid candidates to host the ET observatory: the Sardinia site, close to the Sos Enattos mine, the Euregio Rhine-Meusse site, close to the Netherland-Belgium-Germany border (a third option in Saxony, Germany, is under discussion). Both sites are undergoing a thorough characterization to determine their possibility to host ET, also in terms of logistics and funding possibilities. The choice of the site is expected to happen at some point during 2024.

3.1 Euregio Meuse-Rhine

In this site a 250 m deep borehole has been excavated and equipped, seismic data are under acquisition and analysis and a set of other boreholes are under excavation. An extensive active and passive site characterisation with sensor arrays was carried on in 2021 showing good seismic noise attenuation given by the particular geological structure. The characterisation funded through Interreg grants, and a large proposal for qualifying the site has been essentially approved by the Dutch government.

3.2 Sos Enattos

The Sardinia site is close to the Sos Enattos mine. A long standing characterization of the mine in one of the corners has started some years ago and is still continuing: seismic, magnetic and acoustic noise characterisations are ongoing at different depth in the mine. Furthermore an underground laboratory (SarGrav) has been recently inaugurated. Two 290 m boreholes have been excavated, equipped and data taking is ongoing, while another set of boreholes is expected in 2022 or beginning of 2023. For this site an intense and international surface investigations program is ongoing, its characterisation is funded on regional and national funds, while a large proposal for technology development and engineering design has been submitted to the Italian government.

Conclusions

The Einstein Telescope is a huge enterprise, with a potentially dramatic increase of GW Scientific revenue. The interest of large community and the actions of a reduced set of scientists pushed it through more than a decade, and in the last few years ET acquired a large momentum and now it is a global scale project: on June 2022 the ET collaboration was estabilished with more than 1200 members from 80 Research Units. The sinergy with the European authorities at government level is being achieved, and funds are arriving. The active collaboration with the other third generation project (the US Cosmic Explorer) is a key factor, but

ET can produce great science also in stand-alone configuration Multi-messenger Astronomy with ET will be greatly enhanced, possibly a prioritazion of triggers will be needed due the big number of GW signals.

References

[1] B. P. Abbott *et al.*, (LIGO Scientific Collaboration and Virgo Collaboration), Phys. Rev. Lett. **116**, 061102 (2016)

[2] B. P. Abbott *et al.*, ApJL **848**, L12 (2017)

[3] M. Punturo *et al.*, Class. Quantum Grav. **27**, 194002 (2010)

[4] e.g. ESF Exploratory Workshop: Toward a 3rd generation European Gravitational Wave Observatory - Perugia (2005), http://esf-gw.pg.infn.it/

[5] M. Maggiore *et al.* JCAP03 **050** (2020)

[6] ET steering committee, *ET design report update 2020*, (ET Technical Documentation System), ET-0007B-20 https://apps.et-gw.eu/tds/ql/?c=15418

UHECR results of combined analyses of TA and Auger experiments

A. di Matteo, *L.* Anchordoqui, *T.* Bister, *R.* de Almeida, *O.* Deligny, *L.* Deval, *G.* Farrar, *U.* Giaccari, *G.* Golup, *R.* Higuchi, *J.* Kim, *M.* Kuznetsov, *I.* Mariş, *G.* Rubtsov, *P.* Tinyakov, and *F.* Urban for the Pierre Auger* and Telescope Array collaborations

Abstract. The origin of ultra-high-energy cosmic rays (UHECRs) is still unknown. Their sources are believed to be within the local universe (a few hundred megaparsecs), but deflections by intergalactic and Galactic magnetic fields prevent us from straightforwardly associating UHECRs to their sources based on their arrival directions, making their angular distribution mostly isotropic. At higher energies, the number of potential source candidates and the magnetic deflections are both expected to be smaller, but so is the available amount of statistics. Hence, it is interesting to perform searches for anisotropies using several different energy thresholds. With a threshold of 8 EeV a dipole modulation has been discovered, and with higher thresholds evidence is mounting for correlations with certain nearby galaxies. Neither of the two main UHECR detectors, the Pierre Auger Observatory and the Telescope Array project, has full-sky coverage. Full-sky searches require combining the datasets of both, and a working group with members of both collaborations has been tasked with this. We present an overview of the challenges encountered in such analyses, recent results from the working group, possible ways of interpreting them, and an outlook for the near future.

1 Introduction

1.1 Ultra-high-energy cosmic rays (UHECRs)

Particles with energies $E \geq 1\,\text{EeV} = 10^{18}\,\text{eV} \approx 0.16\,\text{J}$ are known as ultra-high-energy cosmic rays (UHECRs). UHECRs can be detected by huge arrays of particle detectors on the ground, and they are electrically charged (atomic nuclei, mostly protons); as a result, they are deflected by intergalactic and Galactic magnetic fields by $O(30Z(E/10\,\text{EeV})^{-1})^\circ$, where Z is their electric charge, and do not directly point back to their sources. Their arrival directions are nearly isotropically distributed over the full sky: it took 32,000 events for the first anisotropy to be detected with $\geq 5\sigma$ significance (a 6.5% dipole[1] at $E \geq 8\,\text{EeV}$ [2]). It is still not known where or how UHECRs achieve such energies. At the highest energies, their propagation is limited to distances $O(10^2\,\text{Mpc})$ by interactions with cosmic background photons.

*e-mail: spokespersons@auger.org; full author list: https://auger.org/archive/authors_2022_09.html

[1] As of last update [1] using 44,000 events, the estimate of the dipole magnitude has been revised to 7.3% and its significance has increased to 6.6σ.

1.2 The main UHECR detector arrays: Auger and TA

The largest cosmic-ray detector arrays in the world are the Pierre Auger Observatory ("Auger") in Argentina and the Telescope Array (TA) in Utah, USA. Neither TA alone nor Auger alone covers the full sky: TA can only observe the full northern celestial hemisphere and the parts of the southern one closest to the equator (declinations $\delta > -15.7°$), and Auger vice versa (declinations $\delta < +44.8°$). Hence, we can get full-sky coverage by combining the two, and furthermore their fields of view overlap in a band surrounding the celestial equator.

Several joint working groups of members from both the Auger and the TA collaboration have been established since the early 2010s to perform full-sky studies of the energy spectrum, mass composition and arrival directions of UHECRs, as well as direct calibrations of Auger SD stations at the TA site. Certain working groups also include members from other collaborations, such as that on hadronic interactions and shower physics (with EAS-MSU, IceCube, KASCADE-Grande, NEVOD-DECOR, SUGAR and Yakutsk members) and that on neutrino–UHECR angular correlations (with ANTARES and IceCube members). The results from the joint working groups, mainly presented at the International Symposium on Ultra-High-Energy Cosmic Rays (UHECR) and the International Cosmic Ray Conference (ICRC), are listed at http://tiny.cc/Auger-TA. The data used for the latest Auger–TA joint analyses on arrival directions, presented at ICRC 2021 [3, 4], comprise events detected by Auger in the 17 yr period from 1 January 2004 to 31 December 2020 (with an effective exposure of 123,200 km^2 yr sr and 39,157 events with $E \geq 8.57$ EeV, of which 2,625 with $E \geq 32$ EeV) and by TA in the 11 yr period from 11 May 2008 to 10 May 2019 (with an effective exposure of 13,700 km^2 yr sr and 4,801 events with $E \geq 10$ EeV, of which 315 with $E \geq 40.8$ EeV).

1.3 The issue of the cross-calibration of the energy scales

UHECR energy measurements are affected by sizable systematic uncertainties (±14% for Auger, ±21% for TA). If not corrected, a mismatch between energy scales can yield a spurious dipole. (For example, if events with $E_{\mathrm{true}} = 10$ EeV are reconstructed as $E_{\mathrm{rec}} = 9$ EeV by Auger and as $E_{\mathrm{rec}} = 11$ EeV by TA, and we collect all events with $E_{\mathrm{rec}} \geq 10$ EeV, then events with $E_{\mathrm{true}} = 10$ EeV are included if detected by TA but not if detected by Auger, which would appear as if the UHECR flux from the north were larger than from the south.) Hence, possible mismatches in the energy scales should be corrected to the best of our abilities. Measurements in the common declination band can be used for this purpose.

At ICRC 2021 [3], we simultaneously fitted a twice-broken power-law model for the energy spectrum and a power-law model for the $E_{\mathrm{Auger}} \leftrightarrow E_{\mathrm{TA}}$ conversion to the data in a common declination band, obtaining

$$\frac{E_{\mathrm{Auger}}}{10\,\mathrm{EeV}} = 0.857\left(\frac{E_{\mathrm{TA}}}{10\,\mathrm{EeV}}\right)^{0.937}, \qquad \frac{E_{\mathrm{TA}}}{10\,\mathrm{EeV}} = 1.179\left(\frac{E_{\mathrm{Auger}}}{10\,\mathrm{EeV}}\right)^{1.067}.$$

Note that in this fit we used datasets optimized for anisotropy studies at $E_{\mathrm{TA}} \geq 10$ EeV, hence the result must not be extrapolated to lower energies or used outside of the scope of this analysis.

2 Latest results (shown at ICRC 2021)

2.1 Large-scale anisotropies: dipoles and quadrupoles

The flux Φ of UHECRs as a function of the arrival direction $\hat{\mathbf{n}}$ can be expanded into spherical harmonics as $\Phi(\hat{\mathbf{n}}) = \sum_{\ell=0}^{+\infty}\sum_{m=-\ell}^{+\ell} a_{\ell m} Y_{\ell m}(\hat{\mathbf{n}}) = \Phi_{\mathrm{avg}} \times \left(1 + \mathbf{d}\cdot\hat{\mathbf{n}} + \frac{1}{2}\hat{\mathbf{n}}\cdot\mathsf{Q}\hat{\mathbf{n}} + \cdots\right)$, where

Table 1. Our measurements of the dipole and quadrupole moments [3]. The first uncertainty is statistical, the second is due to the uncertainty in the cross-calibration of energy scales. Values in *italics* (**bold**) are locally significant at $\geq 2\sigma$ ($\geq 4\sigma$).

energies (Auger)	[8.57 EeV, 16 EeV)	[16 EeV, 32 EeV)	[32 EeV, $+\infty$)
energies (TA)	[10 EeV, 19.47 EeV)	[19.47 EeV, 40.8 EeV)	[40.8 EeV, $+\infty$)
d_x [%]	$-0.7 \pm 1.1 \pm 0.0$	$+1.6 \pm 2.0 \pm 0.0$	$-5.3 \pm 3.9 \pm 0.1$
d_y [%]	$\boldsymbol{+4.8 \pm 1.1 \pm 0.0}$	$+3.9 \pm 1.9 \pm 0.1$	$+9.7 \pm 3.7 \pm 0.0$
d_z [%]	$-3.3 \pm 1.4 \pm 1.3$	$-6.0 \pm 2.4 \pm 1.3$	$+3.4 \pm 4.7 \pm 3.6$
$Q_{xx} - Q_{yy}$ [%]	$-5.1 \pm 4.8 \pm 0.0$	$+13.6 \pm 8.3 \pm 0.0$	$+43 \pm 16 \pm 0$
Q_{xz} [%]	$-3.9 \pm 2.9 \pm 0.1$	$+5.4 \pm 5.1 \pm 0.0$	$+5 \pm 11 \pm 0$
Q_{yz} [%]	$-4.9 \pm 2.9 \pm 0.0$	$-9.6 \pm 5.0 \pm 0.1$	$+11.9 \pm 9.8 \pm 0.2$
Q_{zz} [%]	$+0.5 \pm 3.3 \pm 1.7$	$+5.2 \pm 5.8 \pm 1.7$	$+20 \pm 11 \pm 5$
Q_{xy} [%]	$+2.2 \pm 2.4 \pm 0.0$	$+0.2 \pm 4.2 \pm 0.1$	$+4.5 \pm 8.1 \pm 0.1$

coefficients with small ℓ correspond to large-scale ($\sim 180°/\ell$) anisotropies and vice versa. The dipole moment **d** is given by $\sqrt{3}(a_{11}\hat{\mathbf{x}} + a_{1-1}\hat{\mathbf{y}} + a_{10}\hat{\mathbf{z}})/a_{00}$, and likewise the components of the quadrupole moment Q are combinations of a_{2m}/a_{00} ($m = -2, -1, 0, +1, +2$).

The dipole amplitude $|\mathbf{d}|$ and the quadrupole amplitude $|\mathsf{Q}|$ are relatively insensitive to magnetic fields, providing some information about sources: coherent deflections mainly affect the directions of **d**, Q rather than their amplitudes, and turbulent deflections only attenuate the 2^ℓ-upole by a factor $O\left(\exp(-\ell^2\Delta\theta^2_{\mathrm{turb}}/2)\right)$, so that most of the $|\mathbf{d}|$ and a sizable fraction of the $|\mathsf{Q}|$ should survive (see also ref. [5]).

With full-sky coverage it is possible to measure the first two multipoles a_{1m} and , a_{2m} in a way that is unbiased regardless of any assumption made about higher multipoles [3]. The results are listed in table 1.

We do not find any deviations from isotropy other than a weakly energy-dependent dipole towards a direction far away from the Galactic Center and a hint of a quadrupole at the highest energies roughly along the Supergalactic Plane. The reconstructed strength of large-scale anisotropies is at the low edge of the range of model expectations such as those of refs. [6, 7], suggesting a medium to heavy mass composition.

2.2 Medium-scale anisotropies: correlations with nearby galaxies

Unlike large-scale anisotropies, in order to search for medium-scale anisotropies we need to focus on the highest energies, where magnetic deflections are expected to be smaller, though this severely reduces the amount of statistics available, making "blind" searches very unlikely to yield strongly significant results. Hence, we performed targeted searches based on two different catalogs, one comprising all types of galaxies at distances between 1 and 250 Mpc, and one comprising starburst galaxies at distances between 1 and 130 Mpc, described in ref. [4] and refs. therein. The test statistic (TS) we used is twice the log-likelihood ratio between a model consisting of an isotropic background plus a weighted sum of Fisher distributions and the isotropic null hypothesis, scanned over the energy threshold $E_{\min}$, the angular scale ψ, and the signal fraction f.

The most significant correlation we found is TS = 27.2 with the starburst galaxy catalog, using $E_{\min}$ = 38 EeV on the Auger scale (49 EeV on the TA scale), $\psi = 15.5°^{+5.3°}_{-3.2°}$, and $f = 11.8\%^{+5.0\%}_{-3.1\%}$; the observed flux and the model prediction are compared in figure 1. The post-trial significance of this is 4.2σ. We also found a correlation TS = 16.2 with the general galaxy catalog, using $E_{\min}$ = 41 EeV on the Auger scale (53 EeV on the TA scale), $\psi = 24°^{+13°}_{-8°}$, and $f = 38\%^{+28\%}_{-14\%}$, though the post-trial significance of this is only 2.9σ.

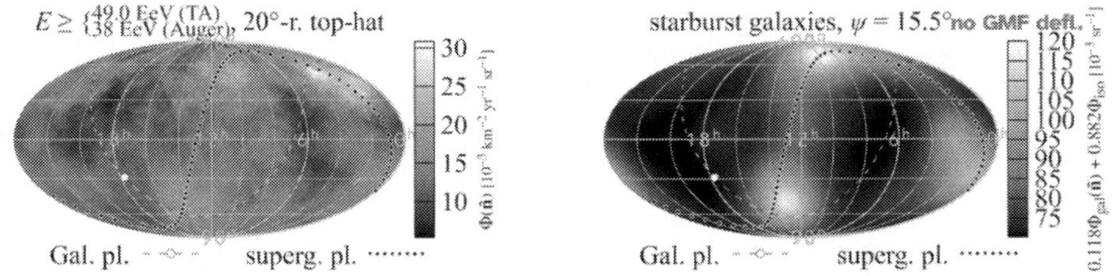

Figure 1. The observed UHECR flux above 38 EeV, compared to the starburst galaxy model predictions

3 Outlook

3.1 Estimating propagation effects on correlation searches via simulations

In order to reduce statistical penalties, the TS was based on a simple model, not taking into account the energy losses of UHECRs (which depend on their mass composition), coherent magnetic deflections, and the possibility of several anisotropic classes of sources at once. In order to estimate their effects, we plan to generate simulated datasets based on a variety of scenarios, subject each of them to the same analyses as the real data at ICRC 2021, and look at which simulations result in similar ψ, f, TS as the real data. The result of such studies will be presented at future conferences.

3.2 Extended datasets

An update of these analyses using three more years of TA data was presented at UHECR 2022 [8], and a further update will be presented at ICRC 2023 using as much Auger and TA data as will be available by then, which we can expect to reduce statistical uncertainties by around 10% compared to those shown here. Continued work by the spectrum working group might reduce uncertainties in the energy cross-calibration even more than this.

3.3 Next-generation experiments and mass-dependent anisotropies

TA is undergoing an upgrade (TA×4) which will increase its area by a factor of 4, helping reduce statistical uncertainties in the northern hemisphere. Auger is undergoing an upgrade (AugerPrime) which will add new scintillation and radio detectors to the existing water-Cherenkov and fluorescence detectors, reducing statistical and systematic uncertainties on UHECR masses. Better UHECR mass estimates will help us study mass-dependent anisotropies, potentially allowing us to disentangle the effects of magnetic deflections from the distribution of UHECR sources. In the farther future, new experiments such as GRAND, POEMMA and GCOS are hopefully going to gather even more data.

References

[1] R. de Almeida et al. (Pierre Auger), PoS **ICRC2021**, 335 (2021)
[2] A. Aab et al. (Pierre Auger), Science **357**, 1266 (2017), 1709.07321
[3] P. Tinyakov et al. (Telescope Array, Pierre Auger), PoS **ICRC2021**, 375 (2021)
[4] A. di Matteo et al. (Telescope Array, Pierre Auger), PoS **ICRC2021**, 308 (2021)
[5] B. Eichmann, T. Winchen, JCAP **04**, 047 (2020), 2001.01530
[6] A. di Matteo, P. Tinyakov, Mon. Not. Roy. Astron. Soc. **476**, 715 (2018), 1706.02534
[7] C. Ding, N. Globus, G.R. Farrar, Astrophys. J. Lett. **913**, L13 (2021), 2101.04564
[8] F. Urban et al. (Pierre Auger, Telescope Array), to be submitted to EPJ Web Conf. (2022)

Astrophysical interpretation of Pierre Auger Observatory measurements of the UHECR energy spectrum and mass composition

Antonio Condorelli[1,*] *on behalf of* the Pierre Auger[2] Collaboration[**]

[1]Laboratoire de Physique des 2 Infinis Irène Joliot-Curie, CNRS/IN2P3, Université Paris-Saclay, France
[2]Observatorio Pierre Auger, Av. San Martín Norte 304, 5613 Malargüe, Argentina

Abstract. We present a combined fit of a simple astrophysical model of Ultra-high-energy cosmic-ray sources to both the energy spectrum and mass composition data measured by the Pierre Auger Observatory. The astrophysical model we adopted consists of identical sources uniformly distributed in a comoving volume, where nuclei are accelerated with a rigidity-dependent mechanism. The fit has been performed at first for energies above 5 EeV, i.e., the region of the all-particle spectrum above the so-called "ankle" feature, and has then been extended to lower energies in order to include this feature. The fit results and their astrophysical implications in both cases are discussed.

1 Introduction

The Pierre Auger Observatory [1] is the largest facility to detect cosmic rays built so far. It is located in the province of Mendoza, Argentina, and has been in operation since 2004, studying cosmic rays by combining different experimental techniques.

The Pierre Auger Collaboration has studied the features of the all-particle energy spectrum with unprecedented precision. Far from being described by a simple power law, in the highest-energy region, the all-particle cosmic-ray spectrum shows several features [2, 3]: a sharp feature, known as the ankle, is observed at $10^{18.7}$ eV, corresponding to a hardening of the spectrum. A new feature, dubbed the instep, at $10^{19.1}$ eV, could reflect the interplay of light-to-intermediate nuclei. Finally, a suppression of the total flux above $10^{19.7}$ eV may be attributed to energy losses during the propagation of ultra-high energy cosmic rays (UHECRs) [4, 5], to the limited maximum energy the sources can provide to particle acceleration, or possibly to a combination of both effects.

The composition of the UHECRs [6, 7], as estimated from the distributions of the depth of maximum development of the showers $X_{\rm max}$, appears to be given from a mix of protons and medium-mass nuclei (e.g. nitrogen) at energies above the second knee, gradually getting lighter with increasing energy up to $10^{18.3}$ eV. The total UHECR spectrum may consist of the superposition of alternating groups of elements with progressively heavier mass, each with a steep cutoff, but with increasingly sparse statistics towards the suppression region. This is

*e-mail: condorelli@ijclab.in2p3.fr
**e-mail: spokespersons@auger.org
A complete list of authors can be found at:
https://www.auger.org/archive/authors_2022_09.html

supported by the observation that the mass composition above the ankle appears heavier and less mixed. Despite the accurate measurement of the energy spectrum, an interpretation in terms of sources is ambiguous. Using both the spectrum and composition, one can remove this degeneracy and infer information about the source scenarios which are compatible with data. Several investigations have been done in recent years to interpret the UHECR spectrum and composition. Most of these studies converge to scenarios with sources injecting hard spectra with low rigidity cutoff and mixed composition, even though simplifying assumptions are used, such as uniform source distributions and 1D cosmic ray propagation. The Pierre Auger Collaboration has published a comprehensive study about the astrophysical implications from the combined fit of spectrum and composition data above the ankle [8], discussing in detail the effects of theoretical uncertainties on propagation and interactions in the atmosphere of UHECRs as well as the dependence of the fit parameters on the systematic experimental uncertainties. In this study, we used a scenario in which the sources of UHECRs are of extragalactic origin and accelerate nuclei in electromagnetic processes with a rigidity-dependent maximum energy $E_{\max}(Z) = E_{\max}(p) \cdot Z$, where Z denotes the charge and $E_{\max}(p)$ is the maximum energy for protons.

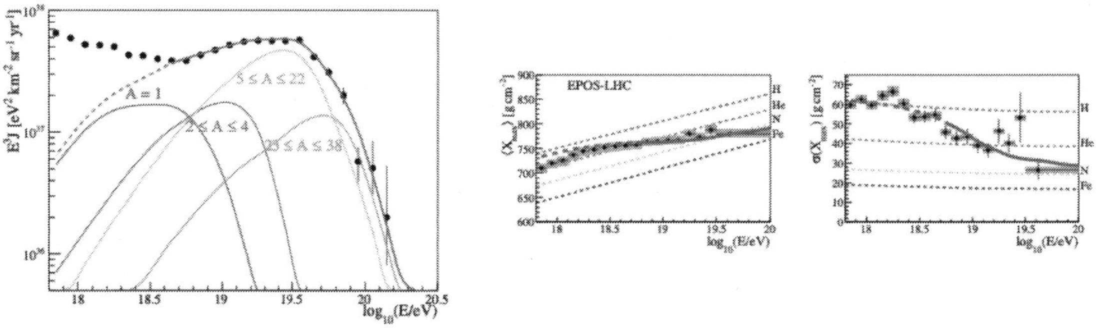

Figure 1. *Left*: Energy spectrum obtained with the best-fit parameters for [8]. Partial spectra are shown with different colors, total spectrum with brown. Full circles show the experimental spectrum. *Right*: average and standard deviation of the $X_{\max}$ for data (full circles) and best-fit prediction (brown). Pure H (red), He (grey), N (green) and Fe (blue) are shown as dashed lines. Only the energy range where the brown lines are solid is included in the fit. From [8].

Within this scenario, a good description of the shape of the measured energy spectrum, as well as the energy evolution of the $X_{\max}$ distributions can be achieved, as shown in Fig. 1, if the sources accelerate a primary nuclear mix consisting of H, He, N and Si. In addition, the primary spectrum has to follow a power law $\propto E^{-\gamma}$ with a spectral index $\gamma \simeq 1$, and the maximum energy of protons has to be about $10^{18.7}$ eV. The mass composition at the sources is dominated by intermediate-mass nuclei (N, Si). Using different hypotheses concerning the extragalactic propagation (more details can be found in [8]), the spectral index and the maximum energy can change considerably and, for some of them, well outside the statistical uncertainties of the fit with other assumptions. However, the hard spectral index and low rigidity cutoff solution are generally preferred.

2 The extended combined fit

In order to provide an astrophysical interpretation of the ankle feature, the fit was extended to lower energies [9]. The simplest extension of this sort consists of introducing a supplemental extragalactic component at low energy, characterized by different physical parameters with respect to the one being dominant above the ankle: from an astrophysical point of view, such a component may originate from a different population of sources. Combining the two contributions, it is possible to have a description of the data including the ankle feature, as shown in the left panel of Fig. 2. The high-energy extragalactic component is qualitatively in agreement with the one found in [8] when fitting only data above $10^{18.7}$ eV, hence the above-ankle fit results are not spoiled by this extension to lower energies.

In this framework, the possibility of including the end of a Galactic contribution at low energies was also explored (right panel of Fig. 2). It was found that a Galactic component made of heavy elements is strongly disfavoured by the composition data, while the best fit was found attributing intermediate-mass to the Galactic spectrum. The presence of a Galactic intermediate-mass contribution, in this case represented by nitrogen, is difficult to accommodate with respect to the standard acceleration in our Galaxy. Nonetheless, such a scenario could be explained by considering an additional Galactic contribution provided by the Wolf-Rayet stars [10], which could actually accelerate nitrogen nuclei up to energies of the order of 10^{18} eV. A detailed discussion of both astrophysical scenarios can be found in [9].

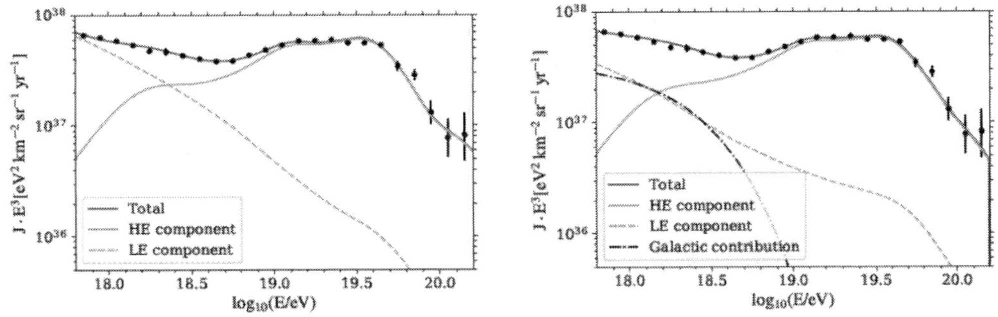

Figure 2. *Left*: The expected low-energy and high-energy contributions in the assumption of pure extragalactic contribution, here shown as dashed and solid lines, respectively. *Right*: The expected low-energy and high-energy contributions, as well as the expected end of the Galactic spectrum, in the assumption that this influences the cosmic-ray flux in the energy region of interest.

3 The sub-ankle proton fit

A complementary approach to extend the combined fit detailed in section 1 was exploited in [11]. For energies between $10^{17.8}$ eV and $10^{18.7}$ eV, only the observed fluxes of protons are fitted, as shown with red dots in the left plot of Fig. 3, while at energies above $10^{18.7}$ eV the procedure is identical to [8]. It was found that including the proton component at low energies significantly worsens the deviance, while considering an additional parameter in the procedure, i.e. the spectral index for protons, the reduced chi-square is compatible with the case exploited in [8]. The best-fit energy spectrum, as well as the experimental data, are illustrated in Fig. 3, while the best-fit parameters are listed in Table 1 of [11].

The fit simultaneously requires a soft spectral index for protons and a hard injection spectrum for intermediate and heavy nuclei at the highest energies. Remarkably, such behavior qualitatively corresponds to scenarios of in-source interactions. In this scenario, the photodisintegration process acts as a high-pass filter allowing the highest energy cosmic-ray nuclei to escape unscattered, whereas the lowest energy ones are disintegrated inside the

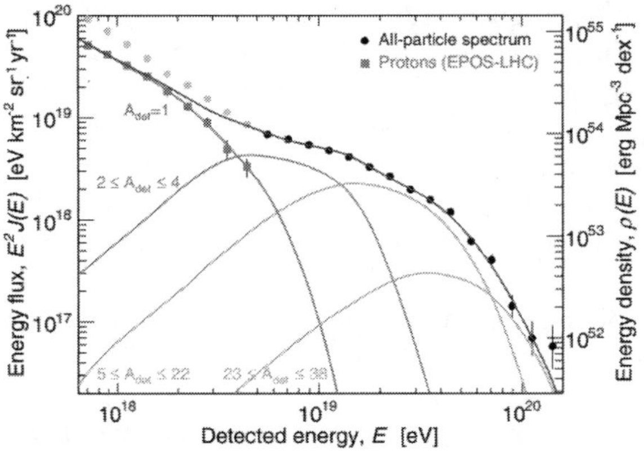

Figure 3. Energy flux at Earth as a function of energy, as modeled by the best-fit parameters for the benchmark scenario. The all-particle spectrum and proton component are shown as black circles and red squares, respectively [11]. Lighter points are not included in the fit. The best-fit components obtained for five detected mass groups are displayed with solid colored lines, as labeled in the Figure. The energy flux of the heaviest mass group, with detected mass number in 39–56, is below the range of interest.

source region, thereby generating a pile-up of nucleons with energy scaling as $1/A$, being A the mass of the nucleus injected in the acceleration region, producing hard ejection spectra for nuclei below the maximum rigidity and generate secondary neutrons at lower energies [12]. These results provide a solid base for constraining the spectral indices for protons and nuclei, as well as the energetics of the sources and the abundances of elements in their surroundings.

Future observations will help us to corroborate the importance of in-source interactions for the interpretation of UHECR data, as shown in this section, or to call for alternative interpretations, such as two extragalactic components from different types of sources, as presented in section 2. For this quest, the upgraded instrumentation of the Pierre Auger Observatory in the next years [13] could play a major role.

References

[1] Aab, A. et al. [Pierre Auger Collaboration] , NIM A **798**, 172-213 (2015)
[2] Abreu, P. et al. [Pierre Auger Collaboration], EPJC **81**, 966 (2021)
[3] Aab, A. et al. [The Pierre Auger Collaboration], Phys. Rev. Lett. **125**, 121106 (2020)
[4] Greisen, K., Phys. Rev. Lett. **16**, 748 (1966)
[5] Zatsepin, G. T. and Kuzmin, V. A., JETP Lett. **4**, 78 (1966)
[6] Aab, A. et al. [The Pierre Auger Collaboration], Phys.Rev. **D90** 122006 (2014)
[7] Bellido, J. and Pierre Auger Collaboration, PoS ICRC2017, 506 (2018)
[8] Aab, A. et al. [The Pierre Auger Collaboration], JCAP**04**, 038 (2017)
[9] Halim, A. et al. [The Pierre Auger Collaboration], arXiv:2211.02857.
[10] Thoudam, S. et al. , A&A **A33**, 595 (2016)
[11] Luce, Q. et al. , *ApJ* **936**, 62 (2022)
[12] Unger, M. et al., Phys. Rev. **D 92**, 123001 (2015)
[13] Castellina, A. et al [Pierre Auger Collaboration] , EPJ Web Conf. **210**, 210 (2019)

Probing Hadronic Interactions with Cosmic Rays

Dennis Soldin[1,2,*]

[1]Karlsruhe Institute of Technology, 76021 Karlsruhe, Germany
[2]University of Delaware, Newark, DE 19716, USA

Abstract. High-energy cosmic rays interact in the Earth's atmosphere and produce extensive air showers (EASs) which can be measured with large detector arrays at the ground. The interpretation of these measurements relies on models of the EAS development which represents a challenge as well as an opportunity to test quantum chromodynamics (QCD) under extreme conditions. The EAS development is driven by hadron-ion collisions under low momentum transfer in the non-perturbative regime of QCD. Under these conditions, hadron production cannot be described using first principles and these interactions cannot be probed with existing collider experiments. Thus, accurate measurements of the EAS development provide a unique probe of multi-particle production in hadronic interactions.

1 Introduction

Throughout the history of elementary particle physics, discoveries have been made through the observation of cosmic rays and neutrinos. Examples are the discovery of new elementary particles, the confirmation of neutrino oscillations, as well as measurements of particle interactions far beyond current collider energies. When a cosmic ray enters the atmosphere and collides with an air nucleus, it initiates particle cascades which form an extensive air shower (EAS). While the decay of neutral pions into photon pairs produces an electromagnetic cascade, charged pions, kaons, and baryons interact again with air nuclei. This process continues until most energy is dissipated through the electromagnetic cascade, and charged particles reach an energy where the decay into muons becomes more likely than (re-)interactions with air nuclei. Thus, muons are generally messengers of the hadronic interactions in EASs.

Experimentally, the properties of the initial cosmic ray, such as energy and mass, are inferred indirectly from the particles measured at the ground by large detector arrays and their interpretation strongly relies on simulations of the EAS development. The main challenge in the description of EASs is the treatment of hadronic interactions in the atmosphere over many orders of magnitude in energy. While interactions of hadrons up to several TeV have been well studied at collider experiments, interactions at higher energies can not be probed by existing facilities. Interactions in the forward direction, which dominate the particle production in EASs, and the combination of particles involved in those interactions can not be measured with current detectors at the Large Hadron Collider (LHC) at CERN. The EAS development is mainly driven by relativistic hadron-ion collisions in the atmosphere at low momentum transfer in the non-perturbative regime of quantum chromodynamics (QCD). Under

*e-mail: soldin@kit.edu

these conditions, hadron production cannot be described using first principles and thus existing simulation packages rely on a variety of phenomenological hadronic interaction models. Commonly used hadronic interaction models are Sibyll [1, 2], QGSJet [3, 4], EPOS [5, 6], and DPMJet [7]. Generally, all these models are based on different realizations of perturbative QCD with Gribov-Regge theory and rely on some fundamental principles, such as conservation laws. However, the particle production is dominated by non-perturbative QCD which is treated by more phenomenological approaches (see e.g., Ref. [8] for a recent review). Large uncertainties remain due to those theoretical limitations and the lack of data from existing collider experiments. Thus, dedicated measurements of hadronic interactions in EASs are crucial in order to test and improve existing hadronic interaction models.

2 Hadronic Interactions and the Muon Puzzle in EASs

Muons are tracers of hadronic interactions and thus the measurement of the number of muons in EASs, N_μ, is an important observable in order to test hadronic interaction models. Over the last 20 years, many experiments reported discrepancies between model predictions and experimental data, while other experiments reported no discrepancies [8]. A systematic meta-analysis of measurements of GeV muons from nine air shower experiments found a global picture of the energy dependence and high statistical significance of the discrepancies [9, 10]. Key aspects of this analysis are a cross-calibration of the energy scales of the different experiments and the definition of the z-value, an abstract measure of the muon content which is comparable between experiments and different analyses. The value $\Delta z = z - z_{\text{mass}}$, where z_{mass} is the number of muons predicted by a hadronic model assuming a mass composition of the primaries based on experimental parameterizations [11], measures the difference between the experimental data and the inferred number of muons for a given hadronic interaction model. A positive value indicates an excess of muons in data with respect to simulations and 0 indicates a perfect match. The resulting distribution of Δz is shown in Fig. 1 for two hadronic models, EPOS-LHC [6] and QGSJet-II.04 [4]. The experimental measurements are consistent within uncertainties with predictions up to EAS energies of approximately 100 PeV. However, above 100 PeV a muon excess is observed that systematically increases with the EAS energy. The slope of a linear fit to this excess is found to be non-zero at $\sim 8\sigma$ for all hadronic interaction models considered. These discrepancies are referred to as the Muon Puzzle and they indicate severe shortcomings in the understanding of particle physics.

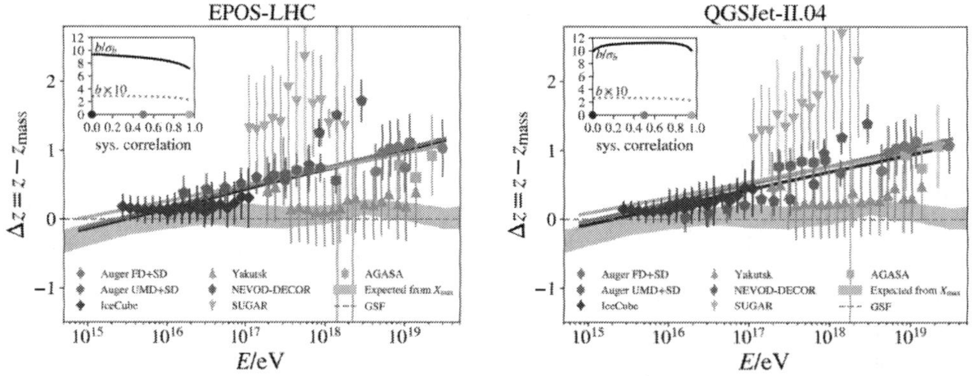

Figure 1. Linear fits of the form $\Delta z = a + b \cdot \log_{10}(E/10^{16}\,\text{eV})$ to the $\Delta z = z - z_{\text{mass}}$ distributions, as described in Ref. [9]. Shown in the inset are the slope, b, and its deviation from zero in standard deviations for an assumed correlation of the uncertainties within each experiment (for details see Ref. [10]). Examples of the fits are shown for a correlation of 0.0, 0.5, and 0.95 in varying shades of gray.

Another important observable to probe hadronic interactions is the standard deviation of the muon number, $\sigma(N_\mu)$, which shows reasonable agreement between experimental data [12] and the models EPOS-LHC [6], QGSJET-II.04 [4], and SIBYLL 2.3D [2]. In addition, discrepancies in measurements of the attenuation length of GeV muons [13] suggest a problem in the description of the energy spectra of muons at production. Preliminary studies also indicate differences in the discrepancies of GeV and TeV muons [14] and inconsistencies in the zenith angle distributions of simulated TeV muons have been reported [15]. The muon production depth in the atmosphere [16, 17] has also been shown to have discrepancies with respect to model predictions, which has been related to a poor description of pion-air interactions in EASs. These observations combined provide strong constraints on changes of certain model parameters that have a large impact on the number of muons in EAS simulations.

3 Constraining Hadronic Interaction Models

Four basic parameters of hadronic interactions are most relevant for the EAS development: the inelastic cross section for hadrons in air, the hadron multiplicity, the elasticity (the energy fraction carried by the most energetic particle), and the ratio of electromagnetic to hadronic energy flow. The impact of modifications to these aspects on air shower observables has been investigated in Ref. [18]. The results for the mean and standard deviation of the muon number, N_μ, and the depth of the shower maximum, $X_{\max}$, for a $10^{19.5}$ eV proton shower, are shown in Fig. 2. The baseline prediction of the SIBYLL 2.1 model [1] was modified with a factor that is 1 for an energy of 10^{15} eV and increases or decreases logarithmically with the beam energy, depending on the value $f(E)$ at some intermediate scale (here $\sqrt{s_{\rm NN}} = 13\,\text{TeV}$). For a thorough description of $f(E)$ and a detailed discussion of the modifications see Refs. [8, 18].

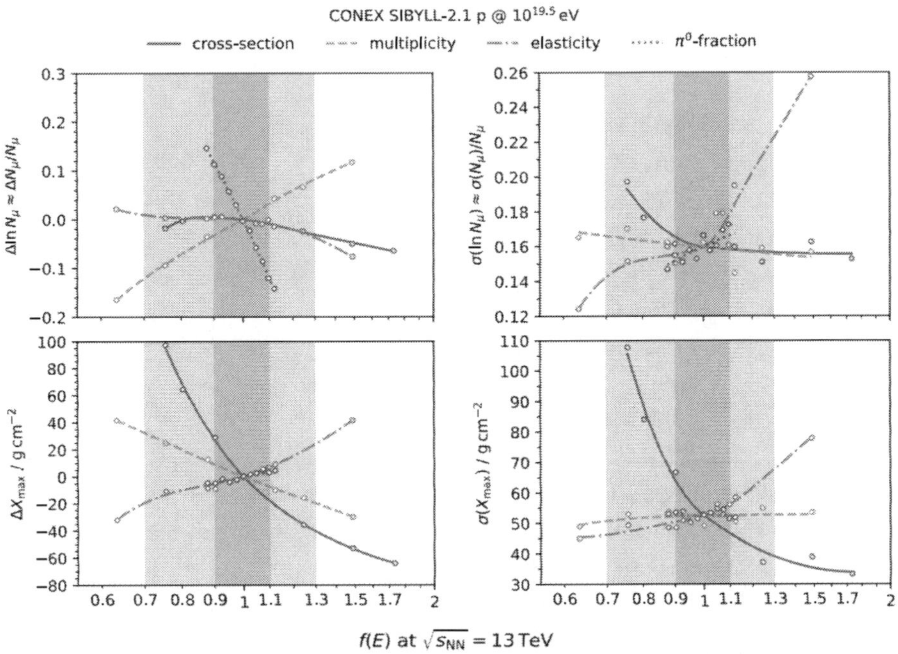

Figure 2. Impact of changing basic parameters of hadronic interactions at $\sqrt{s_{\rm NN}} = 13\,\text{TeV}$ and extrapolating logarithmically (see Ref. [18] for details) on the means and standard deviations of the logarithm of the muon number N_μ (top row) and the depth $X_{\max}$ of the shower maximum (bottom row) for a $10^{19.5}$ eV proton shower simulated with SIBYLL 2.1 [1]. The shaded bands highlight a ±10 % and ±30 % modification. The figure is an update of the original data from Ref. [18], taken from Ref. [8].

As shown in Fig. 2, the most effective way to increase N_μ in EASs is to decrease the π^0-fraction. The number of muons also increases with the multiplicity, but the effect is much weaker. In addition, the observations of the standard deviation of N_μ [12] provides strong constraints on changes to the elasticity, which has a large impact on the muon fluctuations. Since air shower simulations give a reasonable description of the fluctuations, $\sigma(N_\mu)$, and the depth of the shower maximum, $X_{\max}$, it is important to minimize the impact of changes on $\sigma(N_\mu)$, as well as $X_{\max}$ and its standard deviation. This indicates that the only plausible way to increase the muon number sufficiently is to decrease the energy fraction lost to photon production (mainly from π^0-decay) in hadron collisions. There are multiple possibilities to realize a reduced electromagnetic energy fraction within different models which are reviewed in detail in Ref. [8], for example.

4 Conclusions and Outlook

Over the last 20 years, discrepancies between EAS simulations and experimental data have been reported. A first combined meta-analysis of muon data from multiple experiments confirmed an excess of atmospheric muons in EASs above 100 PeV. Dedicated studies of the impact of key parameters in hadronic interaction models indicate that the only plausible way to increase the muon number is to decrease the energy fraction lost to photon production in hadron collisions. However, in order to confirm these studies, further experimental tests are required. Updates of various analyses from existing and new air-shower facilities [19], which include measurements of the muon production depth, the muon energy deposit, and measurements of muons at multiple energies, as well as studies of shower-to-shower fluctuations, will provide crucial information on hadron interactions beyond the phase space of existing colliders to constrain muon production models. In addition, dedicated collider measurements will provide complementary information of hadron production in the forward region [8, 20].

References

[1] E.J. Ahn et al., Phys. Rev. D **80**, 094003 (2009)
[2] F. Riehn et al., Phys. Rev. D **102**, 063002 (2020)
[3] S. Ostapchenko, Phys. Rev. D **74**, 014026 (2006)
[4] S. Ostapchenko, EPJ Web Conf. **208**, 11001 (2019)
[5] K. Werner, F.M. Liu, T. Pierog, Phys. Rev. **C74**, 044902 (2006)
[6] T. Pierog et al., Phys. Rev. C **92**, 034906 (2015)
[7] J. Ranft, Phys. Rev. D **51**, 64 (1995)
[8] J. Albrecht et al., Astrophys. Space Sci. **367**, 27 (2022)
[9] H.P. Dembinski et al., EPJ Web Conf. **210**, 02004 (2019)
[10] D. Soldin et al., PoS **ICRC2021**, 349 (2021)
[11] H.P. Dembinski et al., PoS **ICRC2017**, 533 (2018)
[12] A. Aab et al., Phys. Rev. Lett. **126**, 152002 (2021)
[13] W.D. Apel et al., Astropart. Phys. **95**, 25 (2017)
[14] S. Verpoest, D. Soldin, S. DeRidder, PoS **ICRC2021**, 357 (2021)
[15] D. Soldin et al., EPJ Web Conf. **208**, 08007 (2019)
[16] W.D. Apel et al., Astropart. Phys. **34**, 476 (2011)
[17] A. Aab et al., Phys. Rev. D **90**, 012012 (2014)
[18] R. Ulrich, R. Engel, M. Unger, Phys. Rev. **D83**, 054026 (2011)
[19] A. Coleman et al., Astropart. Phys. **147**, 102794 (2023)
[20] J. Feng et al., J. Phys. G **50**, 030501 (2023)

New approaches for gamma-hadron separation at the IceCube Neutrino Observatory

Federico Bontempo[1,*] for the IceCube collaboration[**]

[1]Karlsruhe Institute of Technology, Institute for Astroparticle Physics, 76021 Karlsruhe, Germany

Abstract. The IceCube Neutrino Observatory located at the geographic South Pole is composed of two detectors. One is the in-ice optical array, which measures high-energy muons from air showers and charged particles produced by the interaction of high-energy neutrinos in the ice. The other is an array of ice-Cherenkov tanks at the surface, called IceTop, which is used both as veto for the in-ice neutrino measurements and for detecting cosmic-ray air showers. In the next decade, the IceCube-Gen2 extension will increase the surface coverage including surface radio antennas and scintillator panels on the footprint of an extended optical array in the ice. The combination of the current surface and in-ice detectors can be exploited for the study of cosmic rays and the search for PeV gamma rays. The in-ice detector measures the high-energy muonic component of air showers, whereas the signal in IceTop is dominated by the electromagnetic component. The relative size of the muonic and electromagnetic components is different for gamma- and hadron-induced air showers. Thus, the gamma-hadron separation of cosmic rays is attempted using machine learning techniques including deep learning. Here, different approaches are presented. Finally, the prospects for the detection of PeV photons with IceCube-Gen2 will be discussed.

1 Introduction

The IceCube Neutrino Observatory [1] at the Amundsen–Scott South Pole Station in Antarctica is both, a neutrino and cosmic-ray detector, made by two arrays: one on the surface and one in the ice. The surface array, called IceTop [2], covers an area of about 1 km^2 and is composed of 81 stations with two ice-Cherenkov tanks per station. The in-ice array occupies a 1 km^3 volume with 86 strings, each with 60 optical modules attached, which measure the Cherenkov light generated by high-energetic particles, such as muons.

The surface detector is constructed to be used as partial veto for the in-ice neutrino detection and for the measurement of cosmic-ray air showers. The veto capability allows to discriminate between muons that are generated in the atmosphere and the ones generated by neutrino interactions. For cosmic-ray studies, the secondary particles, generated by the interaction of the primary with air molecules in the atmosphere, are detected at the surface. IceTop measures primarily the electromagnetic component of the air shower, and through a log-likelihood minimization it is possible to reconstruct the primary information such as

*email: federico.bontempo@kit.edu
**https://icecube.wisc.edu/collaboration/authors/

energy and direction [2]. In addition, muons generated during the shower development with energy above a few hundreds GeV can be detected in the in-ice array.

2 Gamma-hadron separation

Air showers have different developments depending on the primary particle type. Showers with the same energy and direction can show differences in the shower development, for example, in the muon content and lateral spread. In particular, gamma-ray-induced air showers are expected to be almost exclusively electromagnetic thus producing fewer muons and have a narrower lateral spread with less shower-to-shower fluctuations. In contrast, hadronic air showers are richer in muon content and have a wider lateral spread with larger shower-to-shower fluctuations [3]. For the analysis presented here, they will be classified in two categories: gamma-ray or hadronic-induced air showers.

The two categories can be discriminated using the total charge measured in the in-ice detector as an estimator of the high-energy muon component of the air shower. Other parameters taken into account are the reconstructed energy-proxy, S_{125}, as described in the next section, and the reconstructed direction of the primary.

A previous search for PeV gamma rays in IceCube was performed using a random forest classifier [3]. It focused on two main topics: a search for a diffuse flux from the galactic plane and a point source search. The diffuse flux study set an upper limit on the galactic plane gamma-ray emission flux of $2.61 \times 10^{-19} \text{cm}^{-2}\text{s}^{-1}\text{TeV}^{-1}$ at a 90% confidence level. The sky search of gamma-ray point sources was performed after the selection of gamma-ray candidates and was found to be consistent with the background expectation [3].

In this work, a new approach with the same goals is now attempted encouraged by the recent results of the LHAASO collaboration: the first observation of a PeV photon and multiple ultra-high-energy gamma-ray sources [4]. The galactic plane, observed in the Northern Hemisphere, turned out to be rich in multiple TeV to PeV gamma-ray sources. Thus, sources might be expected also in the Southern Hemisphere. For this reason, an estimation of the flux of a potential LHAASO-like source in IceCube's field of view (FOV) is calculated. The flux of three LHAASO sources is extrapolated in the energy range $5.7 < \log_{10}(E/\text{GeV}) < 6.3$ and compared to the expected background, as shown in Figure 1, with an estimated 4° angular resolution for gamma-ray air showers. Four million cosmic-ray events are expected in this energy range per year at IceTop. The number of gamma rays expected in IceTop for a LHAASO-like source depends strongly on the flux assumption. A back of the envelope estimation gives approximately between 10 and 50 events per year. Thus, the expected flux of a LHAASO-like sources in IceCube is roughly 5 orders of magnitude lower than the cosmic-ray background flux.

The discovery of LHAASO-like sources with 10 years of data is expected with a significance of $N_s/\sqrt{N_b} > 5\sigma$, where N_s is the number of photons from the source and N_b is the estimated number of background events. Therefore, the separation between gamma rays and hadrons must be at least between 3 and 4 orders of magnitude. The previous analysis had a background suppression of approximately 2 orders of magnitude and therefore no sufficient sensitivity was reached. Two ways are currently explored for the background suppression. The fist one is an ongoing study to improve the angular resolution, which will provide a better pointing to the potential sources, a smaller opening angle, and consequently, a reduced background. The second way is to discriminate between gamma-ray and hadron-induced air showers via the utilization of machine learning techniques. In this preliminary work, both a random forest regressor and a fully connected neural network are investigated [5]. The random forest regressor is a machine learning technique which consists of a combination of many simple decision trees with a binary splitting being combined together at the end of

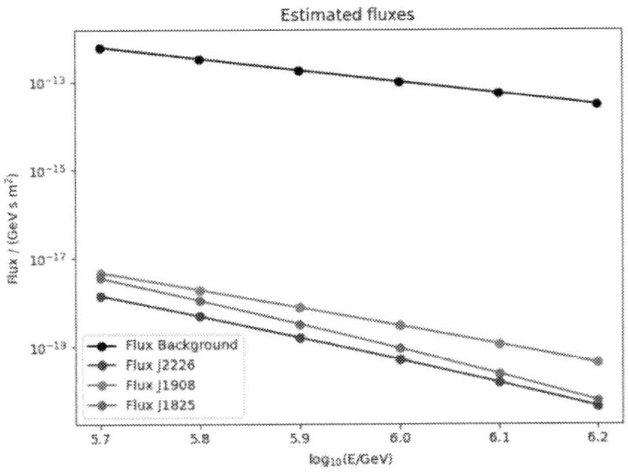

Figure 1: The extended flux of the 3 LHAASO sources [4] up to 2 PeV (colored lines), compared with the background (black line) is shown, an energy range $5.7 < \log_{10}(E/GeV) < 6.3$ and a 4 degree angular resolution for PeV photons at IceCube as circular region in the sky.

training, whereas the deep learning fully connected neural network uses weights and non-linear functions for the same purpose. For the discrimination of signal (gamma rays) and background (hadrons), both models have their output constrained between 0 and 1. Thus, it can be interpreted as a probability for the a posterior gamma-hadron-separation.

Both models are provided with the same input features. Namely, the total in-ice measured charge is used as the indicator for the total high-energy muonic component of the air shower. A containment parameter is utilized for an estimation on how contained the shower axis is in the in-ice detector. The shower size, S_{125} which is the signal expectation at a perpendicular reference distance of 125 meters from the shower axis [6], is taken as the energy estimator parameter. The sine of the declination angle is the indicator of the inclination of the air shower. Finally, a log-likelihood parameter is considered, which estimates how likely it is for an air shower to be induced by a gamma-ray or a hadronic primary [3].

Figure 2 shows the suppression of the background and the loss of signal for the two different approaches for the gamma-hadron separation. Figure 2a shows the performance of the random forest. The passing fraction of events is plotted as a function of S_{125}, the energy proxy, where $\log_{10}(S_{125}) = 0$ corresponds to approximately 1 PeV. The background is given by a small subset of data, which might contain a negligible fraction of gamma rays. In these plots the selection cut is set at 0.999, this means that the all the outputs with values less than the cut value are labeled as background and not passing the selection. Figure 2b uses the same data and input as the random forest, the only difference is that in this case a deep learning fully connected neural network is used for the discrimination. In both plots, the fraction of gamma rays passing the selection cut is approximately constant over the entire energy range. We find that, the deep learning model has a higher passing fraction compared to random forest. However, in order to compare the two methods up to the highest energies, a larger background testing sample is required to enhance the statistics in the relevant region.

3 Conclusion and Outlook

The combination of IceCube's in-ice and surface detectors enables gamma-hadron separation, which can be performed using machine learning methods such as random forest or deep learning. This can be achieved by combining the reconstructed properties of the air shower

(a) (b)

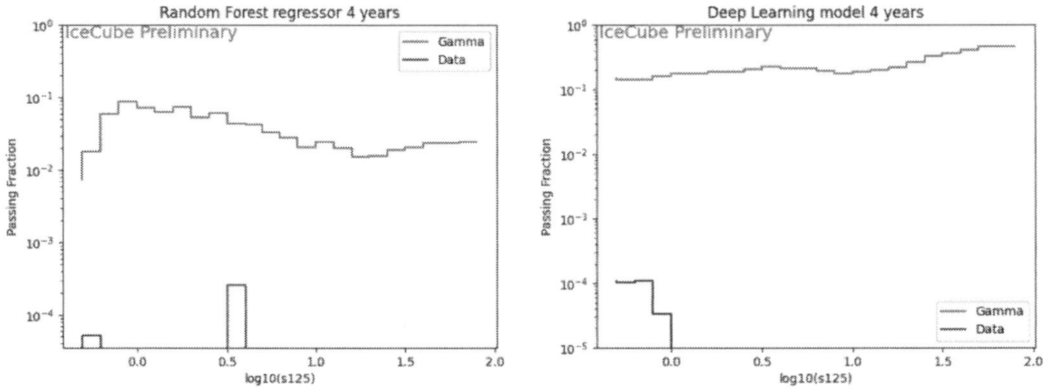

Figure 2: Passing fractions of gamma rays (red line) and data (blue line) as function of S_{125}. The two different models investigated for the gamma-hadron separation are shown. Both use the same input features and have the selection cut set at 0.999. The period of time taken into account is from 2012 to 2015. The missing line in Data means a zero value, obtained on the test sample which counts $\sim$ 10.000 events per year.

such as energy and direction of the primary, the total in-ice charge for the estimation of high-energy muons, and by the log-likelihood parameter. The discovery of gamma-ray sources is expected with 10 years of data available, with a significance of 5σ, if the separation between gamma rays and hadrons is at least between 3 and 4 orders of magnitude.

In the near future, the planned extension of IceCube, called IceCube-Gen2 [7], will have 8 times more statistics for the current field of view (FoV) since it will cover a surface of ~6 km^2 with 150 stations, each composed by 8 scintillator panels and 3 antennas. Together with the increased FoV, the overall aperture will increase by a factor of about 30. This will further increase the discovery potential for gamma-ray sources and the capability for the classification of the primary particle will improve via the combination of ice Cherenkov tanks, scintillator panels, and antennas, providing a better estimation of the muon content of the air showers.

References

[1] M.G. Aartsen et al. (IceCube), JINST **12**, P03012 (2017), 1612.05093
[2] R. Abbasi et al. (IceCube), Nuclear Instruments and Methods in Physics Research Section A: Accelerators, Spectrometers, Detectors and Associated Equipment **700**, 188 (2013)
[3] M.G. Aartsen et al. (IceCube), Astrophys. J. **891**, 9 (2019), 1908.09918
[4] Z. Cao et al. (LHAASO), Nature **594**, 33 (2021)
[5] V. Rodriguez-Galiano, M. Sanchez-Castillo, M. Chica-Olmo, M. Chica-Rivas, Ore Geology Reviews **71**, 804 (2015)
[6] M.G. Aartsen et al. (IceCube), Phys. Rev. D **100**, 082002 (2019), 1906.04317
[7] M. Ackermann et al. (IceCube), Journal of High Energy Astrophysics **36**, 55 (2022)

This project has received funding from the European Research Council (ERC) under the European Union's Horizon 2020 research and innovation programme (grant agreement No 802729).

Detection prospects for multi-GeV neutrinos from collisionally heated GRBs

Angela Zegarelli[1,2,3,*]

[1]Dipartimento di Fisica, Università Sapienza, P.le Aldo Moro 2, I-00185 Rome, Italy
[2]Istituto Nazionale di Fisica Nucleare, Sezione di Roma, P.le Aldo Moro 2, I-00185 Rome, Italy
[3]Dipartimento di Fisica, Università Tor Vergata, Via della Ricerca Scientifica, 00133 Rome, Italy

Abstract. Neutrinos with energies ranging from GeV to sub-TeV are expected to be produced in γ-ray Bursts (GRBs) as a result of the dissipation of the jet kinetic energy through nuclear collisions occurring around or below the photosphere, where the jet is still optically thick to high-energy radiation. So far, neutrino emission from the inelastic collisional model in GRBs has been poorly investigated from the experimental point of view. In the present work, we discuss prospects for identifying neutrinos produced in such collisionally heated GRBs with the large-volume neutrino telescopes KM3NeT and IceCube, including their low-energy extensions, KM3NeT/ORCA and DeepCore, respectively. We evaluated the detection sensitivity for neutrinos from both individual and stacked GRBs, exploring bulk Lorentz factor values ranging from 100 to 600. As a result of our analysis, individual searches appear feasible only for extreme sources, characterized by γ-ray fluence values at the level of $F\gamma \geq 10^{-2}$ erg cm $^{-2}$. In turn, it is possible to detect a significant flux of neutrinos from a stacking sample of $\sim$ 900 long GRBs (which could be detected by current γ-ray satellites in about five years) already with DeepCore and KM3NeT/ORCA. The detection sensitivity increases with the inclusion of data from the high-energy telescopes, IceCube and KM3NeT/ARCA, respectively.

1 Introduction

Gamma-Ray Bursts (GRBs) are the most luminous astrophysical phenomena known to occur in the Universe. These sources emit non-thermal high-energy gamma (γ)-ray emission: the γ-ray energy flux peaks at a few hundred keV and, in many bursts, a long tail extending occasionally up to GeV is present. When the detector's energy sensitivity is wide enough, a typical GRB spectrum can be fitted with a smoothly joined broken power law, known as the Band function [1]. In the classical scenario, the observed γ rays are interpreted as result of synchrotron emission from non-thermal electrons accelerated at the shock fronts in the optically thin region of the GRB jet. Nevertheless, some observed GRB spectra are significantly harder at low energies than what is expected in case of synchrotron radiation. As a possible solution, it was hypothesized that GRB spectra include a bright photospheric component which results from strong subphotospheric heating originating from inelastic nucleon-neutron collisions. This so-called *inelastic collision model* for GRB prompt emission naturally predicts a broken power-law γ-ray spectrum through electromagnetic cascades and Coulomb heating [2, 3].

*e-mail: angela.zegarelli@roma1.infn.it

2 Gamma-ray and neutrino production in the inelastic collisional model

The base assumptions of the inelastic collision model are the existence of a dense, hot and neutron-rich central engine, and a baryonic jet, not dominated by magnetic fields. According to this model, the jet composition could include free neutrons, produced by the dissociation of nuclei by γ-ray photons in the inner regions of the disk. These neutrons would decouple from protons at a radius $R_n < R_{ph}$, that is, below the photospheric radius, where the jet is optically thick. The region between R_n and R_{ph} is characterized by inelastic nuclear collisions between protons and neutrons (*pn* collisions), significantly affecting the dynamics of the jet (subphotospheric collision heating). Charged pions produced in these collisions decay producing positrons, which in turn scatter thermal photons by inverse Compton and produce γ-ray emission. The photospheric thermal radiation, released at R_{ph}, is expected to be modified by subphotospheric collisional heating, becoming nonthermal. From *pn* collisions and subsequent $\pi^{\pm}$ decays, also neutrinos would be produced with an energy correlated with the Lorentz factor of the jet (Γ), and with the relative Lorentz factor of the proton ($\Gamma_p \simeq \Gamma$) and neutron ($\Gamma_n \ll \Gamma$) components, $\Gamma_{rel} = \frac{1}{2}\left(\frac{\Gamma}{\Gamma_n} + \frac{\Gamma_n}{\Gamma}\right) \simeq \frac{\Gamma}{2\Gamma_n}$, through

$$E_\nu \approx 0.1\Gamma\, \Gamma_{rel}\, m_p c^2 \rightarrow E_\nu \simeq 100\ \text{GeV}\left(\frac{\Gamma}{500}\right)\left(\frac{\Gamma_{rel}}{2}\right), \tag{1}$$

implying $E_\nu \sim$ 10-100 GeV neutrinos for Lorentz factors $\Gamma \sim 100 - 1000$ [4, 5]. Note that the neutrino energies involved in this model (sub-TeV) are much lower with respect to the ones investigated so far with neutrino searches from GRBs (TeV-PeV energies), assuming the classical internal-shock and/or magnetically dominated jet models. Since from these searches no GRB-neutrino correlation has been found, it is not possible yet to distinguish among the possible leptonic and hadronic nature of radiation from GRB jets. The search of multi-GeV neutrino emission could provide an alternative method to establish the baryonic nature of GRB jets. Furthermore, detecting neutrinos from GRBs at different energies could be crucial to discriminate among the various physical processes proposed for the GRB γ-ray emission. In particular, detecting multi-GeV neutrinos would support the photospheric scenario with collisional heating mechanisms (due to *pn* collisions) operating inside the jet.

3 Current and future low-energy neutrino detectors

To search for sub-TeV neutrinos from GRBs, compact arrays of 3D photomultiplier sensors are needed. At the South Pole, IceCube is complemented with a smaller Cherenkov detector, named DeepCore, characterized by a higher concentration of optical modules, which allows the detection of neutrinos with energies down to 10 GeV. A similar neutrino telescope, KM3NeT-ORCA, is currently under construction in the Northern Hemisphere, off the French Mediterranean coast, with a much denser configuration of optical modules than that of the high-energy detector KM3NeT-ARCA (currently under construction close to Sicily, in Italy), instead designed for neutrino astroparticle physics studies. Even if the primary goal of low-energy neutrino detectors is to investigate intrinsic properties of neutrino particles (i.e., neutrino oscillations), we here exlore the possibility of astroparticle physics studies with both ORCA and DeepCore in the context of GRB analyses. In this contribution, we summarize the results published in [6].

4 Detector performances for GRB detection

We simulated a search performed by considering upgoing muon neutrinos from collisionally heated GRBs, as to explore sensitivity (at trigger level) of KM3NeT (ORCA+ARCA) and DeepCore+IceCube. Synthetic GRB characteristics were considered for such evaluations, as explained in the following.

Neutrino and background flux estimation

The neutrino spectra produced in collisionally heated GRBs were taken from [5] under the following assumptions: emission released at $R_{\rm ph}$, $\Gamma_{\rm rel} = 3$, and $\xi_{\rm N} = 4$, where $\xi_{\rm N}$ represents the ratio between the dissipated isotropic kinetic energy and the isotropic neutrino energy emerging from inelastic nuclear collisions. The neutrino fluence eventually produced in collisionally heated GRBs scales with the γ-ray fluence ($E_\gamma^2\phi_\nu \sim E_\gamma^2\phi_\gamma = F_\gamma$), peaking around the energy indicated in Eq. (1), hence depending on Γ. The background of upgoing neutrinos due to the atmospheric neutrino flux depends on the duration of the burst (namely, on the temporal window during which 90% of the fluence is expected to be released, T_{90}), as well as on the search angular window around the GRB position, defined through the solid angle $\Omega = 2\pi(1-\cos(\theta/2))$, where θ is the plane angle of the search cone around each GRB. Thus, F_γ, T_{90} and θ are parameters to be carefully chosen. In our estimations, F_γ and T_{90} have been assumed to follow the Fermi γ-ray Burst Monitor (GBM) distributions. Both short (SGRB) and long (LGRB) GRBs were considered, separately, as to correctly characterize the two different populations. Concerning the background estimation, we conservatively set the aperture of the search cone to $\theta = 3\theta_{\nu\mu}$, where $\theta_{\nu\mu}$ is the kinematic angle between the incoming neutrino and the emerging muon directions. The latter was calculated at the energy in which each neutrino telescope would observe the maximum number of neutrinos given the model (i.e., for a given Γ). For more details, see [6].

Detection prospects from an individual GRB

To investigate the possibility of detecting neutrinos from an individual collisionally heated GRB, we considered a GRB with high fluence ($F_\gamma \sim 2\times10^{-3}$ erg cm^{-2}) and evaluated neutrino fluences expected on Earth for several values of Lorentz factor, i.e., $\Gamma = 100$, $\Gamma = 300$ and $\Gamma = 600$ (see Fig. 1(a)). Although the GRB considered in this example is characterized by high fluence (comparable to the highest fluence among all GRBs in the Fermi-GBM catalog), we obtained only a small number of signal events observable in each neutrino telescope ($n_s < 1$), even when the low-energy detectors are integrated with the corresponding high-energy ones (see Fig. 1(b) and Tab. 1). To allow an individual GRB to produce $n_s \geq 1$ in at least one of the considered detectors, it has to be characterized by $F_\gamma \geq 10^{-2}$ erg cm^{-2}, i.e. only nearby and very energetic GRBs could do so. Such a fluentic GRB has never been observed to date.

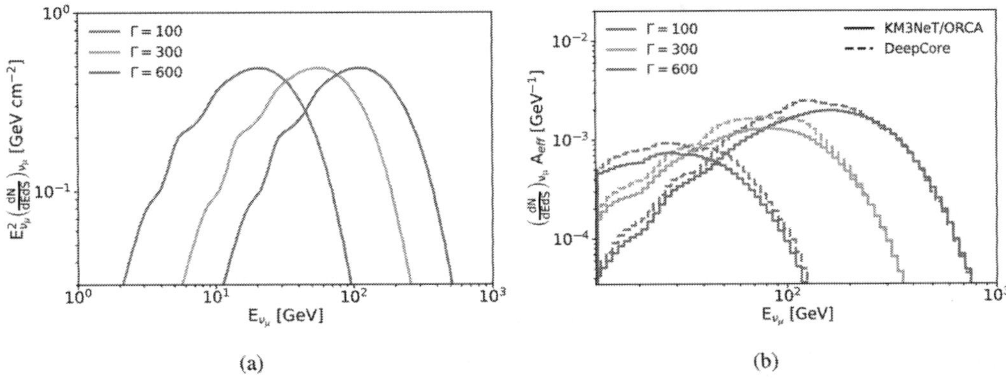

Figure 1: (a) The $\nu_\mu + \bar{\nu}_\mu$ energy fluence for a GRB with observed γ-ray fluence $F_\gamma \sim 2\times10^{-3}$ erg cm^{-2} for Γ=100 (cyan), Γ=300 (orange) and Γ=600 (green). (b) Number of signal events per unit energy in GeV expected in KM3NeT/ORCA (solid lines) and DeepCore (dotted lines) for the neutrino energy fluences shown in (a).

	n_s		
Detector	$\Gamma = 100$	$\Gamma = 300$	$\Gamma = 600$
KM3NeT/ORCA	4×10^{-2}	7×10^{-2}	1×10^{-1}
KM3NeT/ORCA+KM3NeT/ARCA	-	9×10^{-2}	2×10^{-1}
DeepCore	5×10^{-2}	9×10^{-2}	1×10^{-1}
DeepCore+IceCube	-	3×10^{-1}	8×10^{-1}

Table 1: Number of events from $\nu_\mu + \bar{\nu}_\mu$ interactions expected from a GRB with γ-ray fluence $F_\gamma \sim 2\times10^{-3}$ erg cm^{-2} in low-energy detectors (KM3NeT/ORCA and DeepCore) alone, or in a combined search with high-energy detectors (KM3NeT/ARCA and IceCube, respectively). All of these results are given at trigger level.

Stacking detection prospects

The expected detection rate can be greatly increased when summing up the contribution of many GRBs. By taking into account the rate of observed GRBs per year in half of the sky to account for upgoing events only ($N_{SGRB} = 75$ yr^{-1} and $N_{LGRB} = 175$ yr^{-1}), we built a synthetic population of sources where each GRB of the sample is described as explained in Sect. 4. For each extracted source, we estimated the expected neutrino fluence for $\Gamma = [100, 300, 600]$. The background inside the detector is evaluated in a temporal window as wide as $T_{90} \pm 0.3T_{90}$, and in a angular window defined as explained above. Once the GRB sample is defined, we first selected the GRB with the highest level of significance $\sigma = n_s/\sqrt{n_b}$ and, starting from such a GRB, we added one by one the others, choosing each time the GRB that provides the maximum increase of the total level of significance σ_{tot}. We repeated the above procedure 1000 times obtaining at the end the median level of significance and an uncertainty band calculated with percentiles at 1σ and 2σ. By requiring $\sigma_{tot} > 3$ and $n_{s,tot} \geq 1$ as minimum conditions to give a detection, we obtained that for $\Gamma \gtrsim 300$ there is a good chance to detect sub-TeV neutrinos by performing a stacking search of $\sim$ 900 LGRBs with ORCA and DeepCore data, and this possibility is increased if low-energy detectors are integrated with high-energy ones.

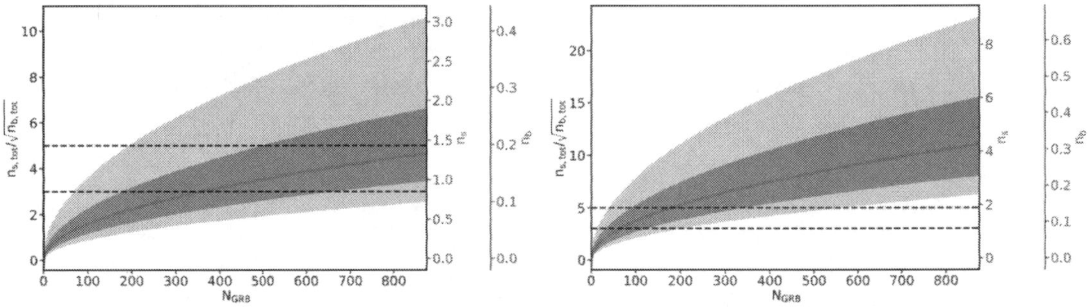

Figure 2: Level of significance $n_{s,tot}/\sqrt{n_{b,tot}}$ for (a) KM3NeT/ORCA+KM3NeT/ARCA and (b) DeepCore+IceCube, achieved by stacking $\sim$ 900 LGRBs with $\Gamma = 300$, within the framework of the inelastic collisional model. The shaded red and gray regions indicate the uncertainty bands at 1σ and 2σ, respectively. The horizontal dashed black lines highlight $\sigma_{tot} = 3$ and $\sigma_{tot} = 5$.

References

[1] Band D. et al., ApJ **413**, 2 (1993)
[2] Beloborodov A. M., MNRAS **407**, 1033 (2010)
[3] Asano K. & Mészáros P., JCAP **09**, 008 (2013)
[4] Bartos I. et al., Phys. Rev. Lett. **110**, 241101 (2013)
[5] Murase K., Kashiyama K., Mészaros P., Phys. Rev. Lett. **111**, 131102 (2013)
[6] Zegarelli A. et al., Phys. Rev. D **105**, 083023 (2022)

Exploiting synergies between neutrino telescopes for the next galactic core-collapse supernova

Meriem Bendahman[1,3], *Anne-Cécile* Buellet[2], *Matteo* Bugli[2], *Joao* Coelho[1], *Alexis* Coleiro[1], *Gwenhaël* de Wasseige[1], *Sonia* El Hedri[1,*], *Thierry* Foglizzo[2], *Davide* Franco[1], *Isabel* Goos[1], *Jérôme* Guilet[2], *Antoine* Kouchner[1], *Yahya* Tayalati[3], *Alessandra* Tonazzo[1], *Cristina* Volpe[1]

[1]Laboratoire Astroparticules et Cosmologie, Université Paris Cité, Paris
[2]CEA Saclay, Department of Astrophysics
[3]Faculty of Sciences, Mohammed V University, Rabat

Abstract. Observing and characterizing the next galactic core-collapse supernova will be a critical step for neutrino experiments. Extracting information about the supernova progenitors and neutrino properties within minutes after an observation will in particular be crucial in order to optimize analysis strategies at other observatories. Moreover, certain classes of progenitors, with strong magnetic fields, could give rise to gamma-ray bursts but have been underinvestigated to date. In this contribution we propose a strategy to combine results from next-generation neutrino experiments, focusing notably on the determination of the progenitor mass and the neutrino mass ordering. Additionally, we investigate the impact of strong magnetic fields on neutrino observations, and demonstrate the detectability of the associated effects in upcoming experiments.

1 Introduction

Observing neutrinos emitted by the next galactic core-collapse supernova (CCSN) would allow a full characterization of its three phases: neutronization, accretion, and cooling. The next generation of neutrino detectors will be uniquely suited to this task: the Hyper-Kamiokande and KM3NeT water Cherenkov (WC) detectors will be sensitive to $\overline{\nu_e}$ while the DUNE and DarkSide-20k liquid argon detectors will be sensitive to ν_e and to the sum of all neutrino flavors, respectively. In this contribution, we will investigate how to use these experiments to uncover novel information about CCSNe and neutrino properties.

2 Combining Water Cherenkov detectors, DUNE, and DarkSide-20k

The time variation of the neutrino rate observed at a given experiment ("light curve") contains crucial information about the CCSN properties and can be obtained less than a minute after the beginning of a supernova. By combining light curves from multiple detectos, alert systems can inform analyses at other observatories. Here, we outline a simple strategy to combine CCSN light curves for detectors sensitive to different neutrino flavors, taking the example of KM3NeT, DUNE, and DarkSide-20k.

*e-mail: elhedri@apc.in2p3.fr

2.1 Multi-detector observables

Existing analyses of CCSN neutrino light curves and spectra are based on supernova simulations. Since accurate three-dimensional (3D) simulations are only available for a limited number of progenitors, over-reliance on simulations could bias our analyses. We therefore elaborated an analysis strategy based on simple assumptions about the flavor complementarity of the neutrino experiments considered, without relying on CCSN simulations. More specifically, we proposed the following two observables:

$$\mathcal{R}(t_{\min}, t_{\max}) = \frac{R_{\mathrm{WC}}(t_{\min}, t_{\max}) + R_{\mathrm{DUNE}}(t_{\min}, t_{\max})}{R_{\mathrm{DS}}(t_{\min}, t_{\max})} \tag{1}$$

$$\mathcal{A}(t_{\min}, t_{\max}) = \frac{R_{\mathrm{WC}}(t_{\min}, t_{\max}) + R_{\mathrm{DUNE}}(t_{\min}, t_{\max}) - R_{\mathrm{DS}}(t_{\min}, t_{\max})}{R_{\mathrm{WC}}(t_{\min}, t_{\max}) + R_{\mathrm{DUNE}}(t_{\min}, t_{\max}) + R_{\mathrm{DS}}(t_{\min}, t_{\max})} \tag{2}$$

where $R_{\mathrm{WC}}, R_{\mathrm{DUNE}}, R_{\mathrm{DS}}(t_{\min}, t_{\max})$ are the neutrino rates observed in the $[t_{\min}, t_{\max}]$ time window at a WC detector, DUNE, and DarkSide-20k, respectively. For a given analysis, the $[t_{\min}, t_{\max}]$ window will be optimized to maximize the search's sensitivity. $\mathcal{R}$ and $\mathcal{A}$ can be interpreted as the ratio and the asymmetry between electronic and non-electronic neutrino flavors. Evaluating the sensitivity of these observables to different supernova parameters will hence allow us to identify which supernova properties warrant a multi-detector study.

2.2 Simple case study with spherically symmetric supernova models

We consider spherically-symmetric supernova simulations from the Garching group [1], which model CCSNe up to the cooling phase, and we assume adiabatic MSW conversions for neutrino flavor transitions. We vary two parameters: the progenitor mass (11 $M_\odot$ vs 27 $M_\odot$), and the neutrino mass ordering. For each configuration we compute $\mathcal{R}$ and $\mathcal{A}$ over different time windows for a supernova occurring at 10 kpc. We choose the fiducial volume of the WC detector to be the one of the KM3NeT experiment, and, for all the detectors considered here, neglect backgrounds and resolution effects. We observe no significant variations of $\mathcal{R}$ and $\mathcal{A}$ with the progenitor mass, with a maximal difference of 1.5σ between the two models. Conversely, as shown in figure 1, the fraction of detected electronic flavors $\mathcal{R}$ in the first 20ms of the supernova is considerably larger in the IMO than in the NMO; measuring $\mathcal{R}$ would in fact allow to determine the neutrino mass ordering at more than 5σ for a galactic supernova. This striking result is explained by the higher ν_e survival probability in the IMO (30%) compared to the NMO (2%) , and should hence hold across a wide range of CCSN models. In the next section, we investigate this result for more realistic supernova models and detector properties.

2.3 Stable observables: from the neutrino mass ordering to new physics

The dependence of $\mathcal{R}$ on the neutrino mass ordering is driven by the variations in the ν_e rates expected at DUNE. Conversely, DarkSide-20k, which treats all flavors equally, provides a reference point to estimate the electron neutrino survival probability independently of the supernova distance. In this section we therefore focus on these two experiments, and use the observable $\mathcal{R}' = R_{\mathrm{DUNE}}/R_{\mathrm{DS}}$ to probe neutrino flavor conversion processes. Specifically, we investigate the variations of $\mathcal{R}'$ with the electron neutrino survival probability p_{ee}, for state-of-the-art 3D supernova models designed by the Garching group [1] with progenitor masses ranging from 15 to 40 $M_\odot$. We also consider realistic detector conditions, using the SNEWPY [2] tool to model detection efficiencies and energy smearing. However, instead of the optimal $[1, 10]$ ms window used in section 2.2, the new simulations constrain us to use a $[11, 20]$ ms window, making our results overpessimistic.

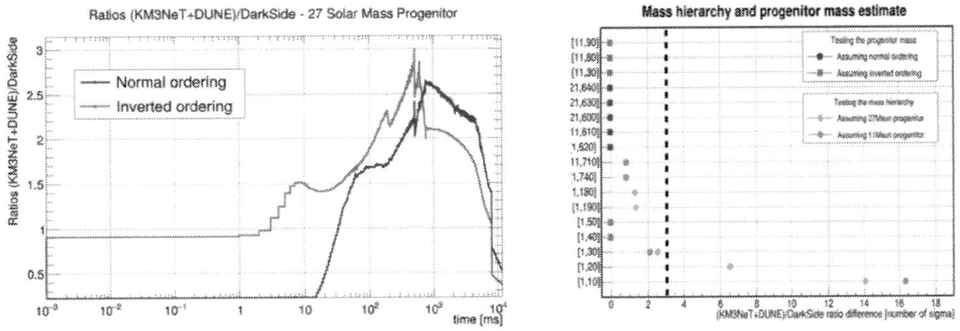

Figure 1. Left: Values of $\mathcal{R}$, computed in 1 ms bins, as a function of time for a 27 $M_\odot$ progenitor. Right: Significances of the differences in $\mathcal{R}$ between models for different time windows.

Figure 2 shows $\mathcal{R}'$ as a function of p_{ee} for the progenitor masses simulated in [1]. The (linear) dependence of $\mathcal{R}'$ in p_{ee} is remarkably stable across models, due to the near-universality of the size of the neutronization peak [3]. While numerous analyses exploiting the size of this peak have been proposed in the literature (see e.g. [3]), a novel aspect of our study is the use of DarkSide-20k to make $\mathcal{R}'$ independent of both the CCSN model and distance. Being exclusively sensitive to neutrino properties, $\mathcal{R}'$ could hence be a unique probe of new physics phenomena. The right panel of figure 2 shows the 1σ absolute uncertainty on the value of p_{ee} obtained from a measurement of $\mathcal{R}'$ as a function of the supernova distance to Earth, for the NMO and the IMO. For a close-by supernova, if the mass ordering is known, a p_{ee} measurement would allow probing new physics scenarios such as neutrino two-body decays, which could modify DUNE's early-time measurements by up to an order of magnitude [4].

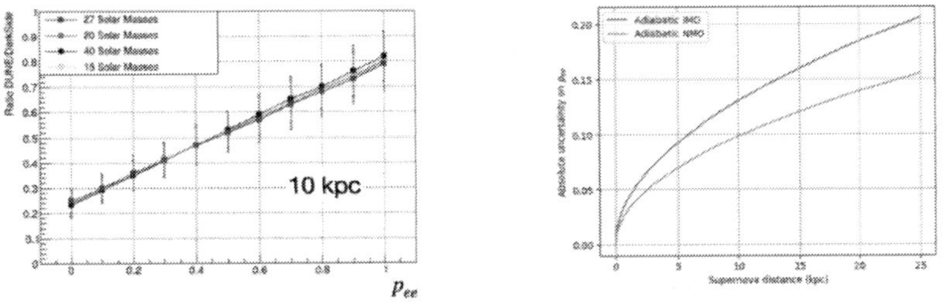

Figure 2. Left: $\mathcal{R}'$ as a function of p_{ee} for a CCSN at 10 kpc. Right: Uncertainty on the measured p_{ee} as a function of the CCSN distance in the NMO and the IMO.

3 Atypical supernovae: progenitor rotation and magnetic fields

Supernova progenitors with strong rotations and magnetic fields can lead to extreme explosions producing long gamma-ray bursts and studying their light curves could therefore advance our understanding of extreme astrophysical phenomena. Here, we investigate the impact of magnetic fields on neutrino emission, using simulations from [5]. These simulations model a $35 M_\odot$ fast-rotating progenitor and an initial 10^{12} G magnetic field with different possible radial profiles and angular momenta. A reference model with no magnetic field is also provided. For all magnetic field configurations, we observed a suppression of the neutrino emission rate in the progenitor's equatorial plane during the accretion phase. This suppression is due to the magnetic field transporting angular momentum towards the periphery of the star, thus inhibiting the matter accretion rate perpendicularly to the rotation axis.

The effect of this flux suppression on neutrino observations is shown in figure 3 for the KM3NeT experiment, assuming the full configuration of the ARCA detector and a CCSN occurring at 5 kpc. Here, in spite of accounting for the detector efficiency and the presence of backgrounds, we observe a clear suppression of the neutrino flux for the $\ell = 1$ model in 200-400 ms. Finally, we assess the observability of this suppression by comparing the numbers of expected events in 200-400 ms for the no-magnetic field model and the $\ell = 1$ model, at different detectors. The right panel of figure 3 shows the significance of this flux suppression for KM3NeT, Hyper-Kamiokande, DUNE, and DarkSide-20k. Even at KM3NeT, the effect of magnetic fields is still observable up to the galactic center at 3σ, while other experiments have full galactic sensitivity. This simple study confirms that magnetic fields can leave detectable imprints on neutrino spectra for a wide range of supernovae. An in-depth investigation involving lighter, more representative progenitors, is therefore warranted.

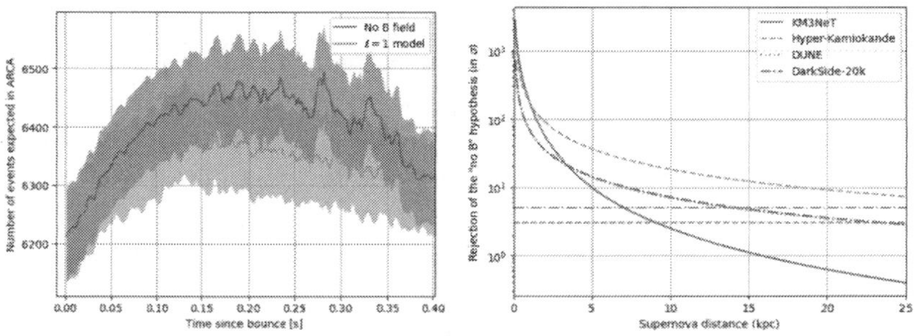

Figure 3. Left: Neutrino rates expected at KM3NeT-ARCA for the rotating progenitors simulated in [5]. Right: Significance of the difference between the "no magnetic field" and the dipolar models as a function of the CCSN distance.

4 Conclusion

In this contribution we presented new approaches to link supernova parameters to neutrino observations. We designed a generic method to identify mechanisms modifying the flavor composition of CCSN neutrinos, and combined experiments to probe new physics phenomena. Furthermore, we showed that the presence of strong magnetic fields could be detected by upcoming neutrino experiments and thus should be targeted by future CCSN analyses.

5 Acknowledgments

LEAK is supported by LabEx UnivEarthS (ANR-10-LABX-0023 and ANR-18-IDEX-0001).

References

[1] https://wwwmpa.mpa-garching.mpg.de/ccsnarchive/archive.html
[2] A. Baxter, *et al.*, J. Open Source Softw. **6**, no.67, 3772 (2021)
[3] M. Kachelriess, *et al*, Phys. Rev. D **71**, 063003 (2005)
[4] A. de Gouvêa, *et al*, Phys. Rev. D **101**, no.4, 043013 (2020)
[5] M. Bugli, *et al*, Mon. Not. Roy. Astron. Soc. **492**, no.1, 58-71 (2020)

Neutrinos from core-collapse supernovae at KM3NeT

Meriem Bendahman[1,2] *on behalf of the KM3NeT Collaboration*
E-mail: mbendahman@km3net.de

[1]Laboratoire Astroparticules et Cosmologie, Université Paris cité, Paris
[2]Faculty of Sciences, Mohammed V University in Rabat, Rabat

Abstract. The SN1987A supernova was the first extragalactic neutrino detection, but no further observations have been made since. Detecting neutrinos from a galactic supernova would provide invaluable information on the supernova mechanism and particle behavior in dense environments, hence, improving the sensitivity of current and upcoming neutrino experiments is crucial. In this contribution, we discuss how the optical module design of the KM3NeT neutrino experiment would allow to observe supernova neutrinos. We present KM3NeT's sensitivity to galactic supernovae and describe its associated online alert system for multi-messenger studies. Finally, we discuss KM3NeT's ability to infer the supernova evolution from the time profile of the associated neutrino emission.

1 Introduction

Thermal neutrinos emitted in the MeV range by supernovae can provide unique information on their explosion phases and can be used to probe neutrino oscillations in dense environments [1]. One way to detect neutrinos from core-collapse supernovae is by using neutrino telescopes, such as KM3NeT which is a large-scale neutrino detector located in the Mediterranean Sea. Although KM3NeT is sensitive to neutrinos of energies range from GeV to PeV. ORCA detector for the detection of neutrinos in the GeV energy range and used to study the neutrino properties and ARCA detector aimed to detect high-energy (TeV to PeV) neutrinos coming from extra-galactic sources [2], KM3NeT can be used for the detection of MeV neutrinos from Supernovae thanks to the structure of its Digital Optical Modules (DOMs). Indeed, these DOMs are composed of multiple photomultipliers (PMTs) allowing a fine-grained analysis of even low-energy signals. We discuss the sensitivity of KM3NeT to CCSN neutrinos and alert system for multi-messenger studies, and its ability to infer supernova evolution from the neutrino emission time profile.

2 Core-collapse Supernovae neutrino detection with KM3NeT

Detecting MeV neutrinos from a CCSN with KM3NeT is a difficult task: these events typically activate only one DOM and are hence extremely difficult to distinguish from ambiant radioactivity and tenuous atmospheric muon signatures. To overcome this challenge, we first removed a large fraction of atmospheric muon signatures by vetoing events associated with triggered multi-DOM patterns. We then devised an observable, the multiplicity, defined as the number of PMTs activated in a given DOM within a 10ns window. The left plot of figure 1 shows the events per a block of 2070 DOMs as a function of the multiplicity for CCSN neutrino signals, and background. In the analysis described in [3], the detection of CCSN neutrinos is based on counting an excess of intermediate multiplicity coincidences over the expected background. The right plot of figure 1 shows the sensitivity as a function of the distance for 3 progenitor models for event multiplicities in the 7-11 range. Taking into account the distribution of CCSNe as a function of the distance to the Earth, more than 95% of the Galactic core-collapse supernovae can be observed by the KM3NeT detectors.

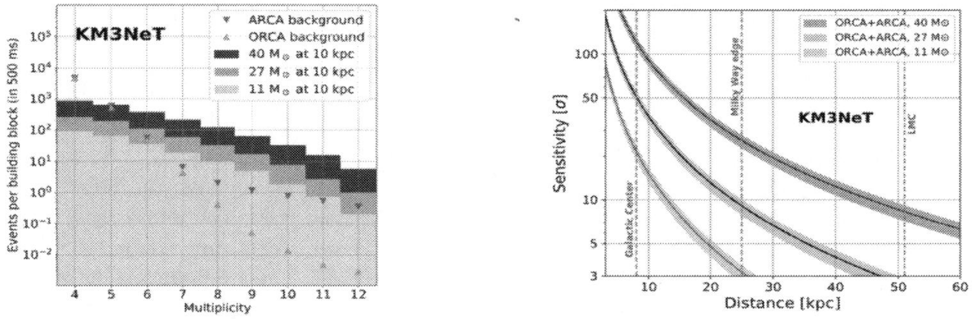

Figure 1. Left: Number of events per building block as a function of the multiplicity for ORCA (light blue) and ARCA (dark blue) for progenitor masses of 11 M⊙ (light orange), 27 M⊙ (orange) and 40 M⊙ (dark orange). Right: KM3NeT detection sensitivity to CCSN neutrinos as a function of the distance to the star for 11 M⊙ (green), 27 M⊙ (gray) and 40 M⊙ progenitors (violet) [3].

3 Supernova characterization

The large event statistics collected by KM3NeT can be exploited to analyse the time profile of CCSN neutrino emission with a potential millisecond time resolution. The analysis of this time profile can provide a tool for model discrimination. In this study, two analyses are considered: the estimation of the neutrino spectrum parameters and the detection of the standing accretion shock instability (SASI).

3.1 Estimation of the neutrino spectrum parameters

CCSN neutrinos have a pinched Fermi-Dirac spectrum characterized by the mean neutrino energy $\langle E \rangle$, the spectral shape parameter α and the signal scale, Λ, which depends on the total energy released and on the distance to the source [3]. In this study, we estimate these parameters by fitting the observed multiplicity distribution, considering the following 3 hypotheses: α free in the full range of expected values 2–4, α in the range $\alpha_{true} \pm 10\%$, and α fixed to the true value. The results of the fit for a given neutrino spectrum are shown on the left panel of figure 2. The left plot of figure 2 shows the signal rate as a function of the mean energy - the signal rate is the number of events scaled by the mean energy and the distance to the source. Results for the uncertainty in the mean neutrino energy are:: ± 0.5 MeV if the spectral index and signal scale are known, below ± 1.5 MeV if the spectral index and signal scale are known within a 10% variation, the sensitivity is lost if the spectral index and signal scale are left free.

3.2 Detection of the Standing accretion shock instability (SASI)

The Standing accretion shock instability (SASI) is a hydrodynamic instability inside the CCSN progenitor which causes fast variations of the neutrino rates give an order of magnitude estimate. Using a model independent approach, which is based on the detection of a significant peak on the power spectrum of the neutrino time profile at any frequency, a 3σ sensitivity to the SASI signature is reached for Galactic progenitors at distances between 3 kpc (27M⊙) and 5 kpc (20 M⊙).

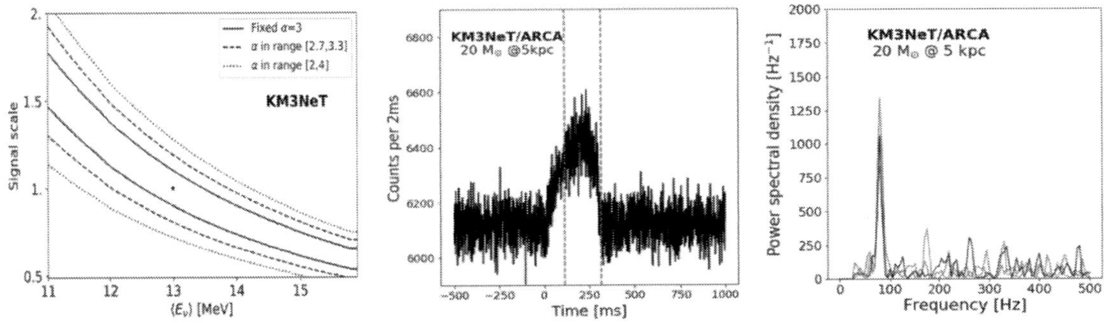

Figure 2. Left: 90% confidence level contours in the signal scale and neutrino energy parameter space. Middle: the neutrino rates in the full ARCA detector for of a 20 M⊙ progenitor at 5 kpc. The dashed red lines correspond to the interval where SASI is searched for [3]. Right: Power spectral densities for ARCA observations. 80 Hz corresponds to the SASI frequency for the 20 M⊙ progenitor at 5 kpc [3].

4 Multi-messenger analysis

4.1 Real-time Multi-Messenger Analysis Framework

Core-collapse supernova neutrinos may be observed hours before the photons, and can be used as an early warning, if a CCSN monitoring is setup. KM3NeT has been active in the framework of the SNEWS network since 2019 [4]. Figure 3 shows the KM3NeT monitoring system for MeV neutrinos.This pipeline is operational and can send alerts with false alarm rate less than one per week to SNEWS. The alert generation latency is less than 20s.

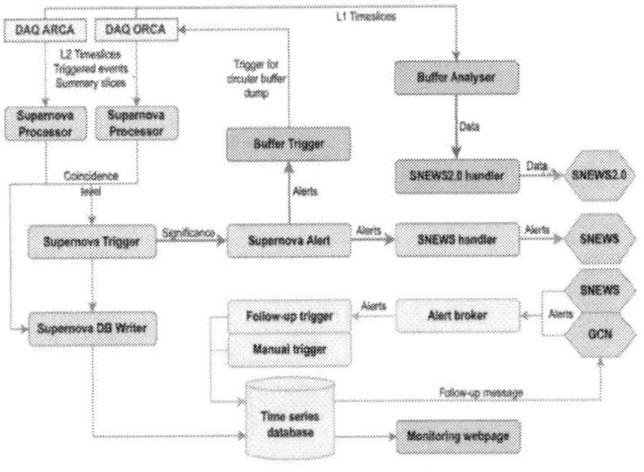

Figure 3. Monitoring the neutrino sky for the next Galactic CCSN with KM3NeT [5].

4.2 Arrival time of the CCSN neutrino signal

The arrival time of the neutrino signal from a core-collapse supernova (CCSN) is expected to occur several hours before the visible light from the explosion reaches Earth. The exact timing of the neutrino signal will depend on the specific properties of the supernova and the distance to the explosion. To extract the arrival time of the first supernova neutrinos, T0, the method described in [3] is used. A moving-average filter with a 23 ms time window is applied. Then, the beginning of the light curve is fitted to an analytical function, with T0 as a parameter. A result of this study is shown on the left plot of figure 4, which presents the

CCSN neutrino time profile of the light curve in ARCA for a 20 $M_{\odot}$ progenitor at a distance of 5 kpc. The arrival time can be estimated with an uncertainty of 3 ms for a supernova at 5 kpc.

4.3 Multi-detector approach for CCSN triangulation

An analysis that uses data from multiple neutrino telescopes has been performed with the objective of locating the CCSN by estimating the time delays between the light curves recorded by different detectors. The right panel of figure 4 shows the 90% confidence area obtained combining measurements with the combination of Hyper Kamiokande-IceCube-JUNO and KM3NeT/ARCA detectors [6]. This confidence area is 140° ± 20°.

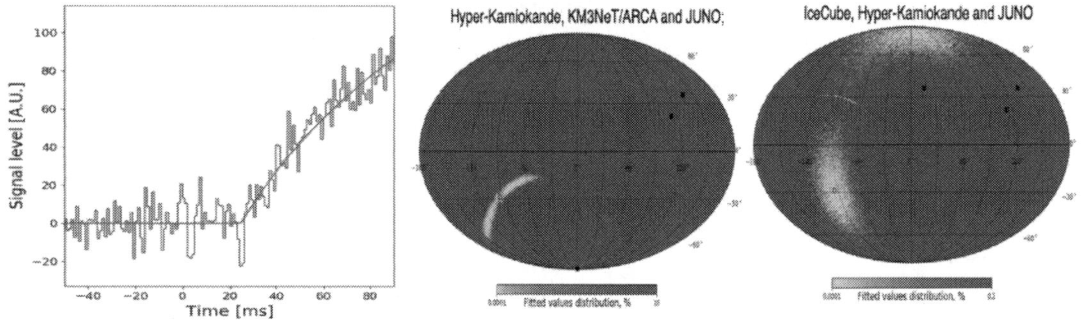

Figure 4. Left: Time profile of the signal in ARCA (2 building blocks) using all coincidences for a 20M progenitor at 5 kpc [3]. Right: Fitted position distributions in equatorial coordinates for a CCSN at the Galactic Centre (black dot) computed using triangulation between different detector combinations [6].

5 Conclusion

In this contribution, KM3NeT's sensitivity to galactic supernovae is presented. More than 95% of the Galactic CCSNe can be detected thanks to the multi-PMT design of the experiment's optical modules. The arrival time of the CCSN neutrino signal can be estimated with an uncertainty of 3 ms for a supernova at 5 kpc. A real-time monitoring pipeline is operational and sending alerts with false alarm rate less than 1/week to SNEWS.

6 Acknowledgments

This work is supported by LabEx UnivEarthS (ANR-10-LABX-0023, ANR-18-IDEX-0001).

References

[1] K. Scholberg, Ann.Rev.Nucl.Part.Sci. 62 (2012).
[2] S. Aiello et al., JINST 17 (2022) P07038.
[3] S. Aiello et al., Eur. Phy. J. C81 (2021) 445.
[4] S. Aiello et al., Eur. Phys J. C82 (2022) 317.
[5] G. Vannoye et. al., https://doi.org/10.5281/zenodo.6785410 (2022).
[6] A. Coleiro et al., Eur. Phys. J. C 80 (2020) 856.

Searching for multi-messenger signals with the Pierre Auger Observatory

Eleonora Guido[1,*] for the Pierre Auger Collaboration[2,**]

[1]Department Physik, Universität Siegen, 57068 Siegen, Germany
[2]Observatorio Pierre Auger, Av. San Martín Norte 304, 5613 Malargüe, Argentina

Abstract. The Pierre Auger Observatory [1], primarily designed for the detection of ultra-high energy (UHE) cosmic rays, has been proven to be also an excellent tool in multi-messenger searches. With its unprecedented exposure to UHE particles, it is exploited to set stringent upper limits on the diffuse flux of UHE photons and neutrinos and to look for neutral particles associated with steady sources and transient events, such as gravitational waves. All these searches can provide key information to investigate the most energetic phenomena in the Universe and answer some of the most important still-open questions in astrophysics.

1 Introduction

Multi-messenger astronomy aims at combining the information from any particle and radiation coming from astrophysical objects to get a deeper insight on their nature. Such studies underwent a significant boost after the first detection of gravitational waves and the rise of neutrino astronomy [2, 3].

The Pierre Auger Observatory, the largest experiment in the world for the detection of ultra-high energy particles, exploits the properties of air showers to identify the ones induced by UHE photons and neutrinos among the hadronic background. Since many models for the origin of UHE cosmic rays predict the production of neutral particles either at the sources or during the propagation through the Universe, the current upper limits have already set strong constraints on them. Moreover, neutral particles directly point back to their production site, which allows their association with specific sources or events. An overview of the on-going multi-messenger activities within the Pierre Auger Collaboration is here presented.

2 Searches for photons

In many models UHE photons originate from the decay of neutral pions produced by the interactions of protons with the Cosmic Microwave Background (CMB) photons and/or with photon fields within the source environment. However, since UHE photons also interact during their propagation and may be eventually absorbed, only distances to about a few tens of kpc can be explored with photons around 10^{15} eV, which rises to few Mpc around 10^{19} eV.

*guido@hep.physik.uni-siegen.de
**spokespersons@auger.org
A complete list of authors can be found at: https://www.auger.org/archive/authors_2022_09.html

The identification of photon-induced air showers relies on the fact that they are dominated by electromagnetic interactions, hence they exhibit a deeper maximum development in the atmosphere $X_{\max}$, which is discriminated with the Fluorescence Detector (FD), and a smaller amount of muons at the ground, which is associated with a slower-rising signal in the single SD stations. Both the deeper $X_{\max}$ and the lower number of muons are reflected in a steeper lateral distribution function (LDF) measured by the Surface Detector (SD) [4]. No UHE photon has been unambiguously detected so far, but Auger set the most stringent upper limits on their flux.

2.1 Diffuse photon flux

Upper limits on the unresolved flux of photons have been set above 2×10^{17} eV. Different techniques are used depending on the energy range.

For the lowest energies the low-energy extensions of the Pierre Auger Observatory are used in a hybrid configuration, i.e. the 750 m SD array and the High Elevation Fluorescence Telescopes (HEAT); three quantities are combined in a multivariate analysis using a Boosted Decision Tree (BDT): the $X_{\max}$ measured by the FD, the number of triggered SD stations and a SD observable sensitive to the difference in LDF called S_b [5]. In the energy range between 10^{18} eV and 10^{19} eV hybrid events are also used in a linear Fisher discriminant analysis, to combine $X_{\max}$ with a parameter F_{μ}, related to the muon content and derived through the air shower universality concept [6]. As concerns energies above 10^{19} eV, only the SD data are used because of the lack of FD statistics and a Fisher discriminant analysis is used to combine two variables, one related to the steepness of the LDF and one to the risetime of the SD signals [7]. The upper limits on the photon flux at 95% confidence level are shown in Fig. 1 in the whole energy range [8].

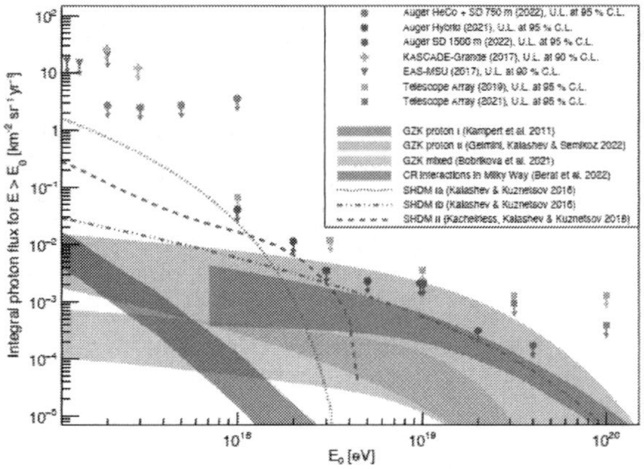

Figure 1. Upper limits on the photon flux at 95% C.L., shown together with predictions from models and other experimental results [8].

2.2 Photons from point-like sources and transient events

Both blind and targeted searches for point-like steady sources of UHE photons are performed in Auger. The analyses are applied to the energy range between $10^{17.3}$ eV and $10^{18.5}$ eV; a BDT is used to select photon-like air showers, with $X_{\max}$ and S_b as the main input variables. No evidence for an excess of photon-like events from any considered direction has been found with the blind search and the corresponding upper limits on the flux of UHE photons are shown in Fig. 2 for each direction [9]. In the case of the targeted search, 12 predefined target classes are chosen, which

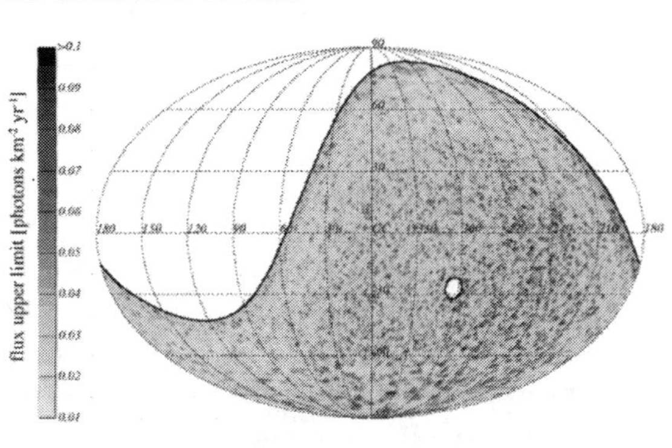

Figure 2. Upper limits on the photon flux as a function of the Galactic coordinates in the celestial map [9].

contain mostly Galactic sources, as well as two extragalactic target sets (three sources of γ-rays in the Large Magellanic Cloud and the core region of Centaurus A) and the Galactic center. No evidence for sources emitting photons in the considered energy range has been found [10].

The Pierre Auger Observatory also performs follow-up searches for photons in association with gravitational waves. Since the UHE photons are significantly attenuated because of the interactions with background radiation fields, the gravitational wave (GW) events are carefully selected to accept only close or well-localized sources; UHE photons are looked for in specific time windows after and around the gravitational wave event. Preliminary results have been obtained by considering a selection of four GW events and no photon candidate was identified, but the first upper limits were set [11].

3 Searches for neutrinos

Similarly to photons, neutrinos at UHE are also produced as a result of the decay of charged pions generated in the interactions of cosmic rays with photon fields. They can travel very long distances without being absorbed and hence are excellent probes into the extragalactic sources and their cosmological evolution.

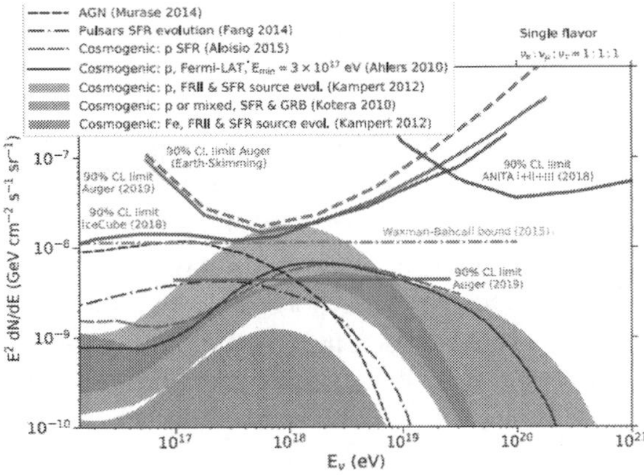

Figure 3. Upper limits at 90% C.L. to the normalization k of the diffuse flux of neutrinos, shown together with predictions from models and other experimental results [12].

Neutrinos, having a greater interaction length with respect to hadrons, interact deeper in atmosphere. As a consequence, inclined showers induced by neutrinos are still "young", i.e. rich in electromagnetic component, when they reach the ground and exhibit broader signals in the SD stations. Hence, inclined showers ($\theta > 60^o$), the ones associated to a very elongated footprint on the SD stations, are used to discriminate neutrinos. In addition to inclined down-going (DG) showers, also Earth-skimming (ES) upward-going showers can be considered ($90^o < \theta < 95^o$): they are associated to the decay in atmosphere of τ leptons, produced by the interactions of ν_τ in the Earth's crust. Even if no UHE neutrino has been unambiguously identified so far, the current upper limits set by Auger are among the most stringent ones.

3.1 Diffuse neutrino flux

Neutrinos with energies above 10^{17} eV can be detected with the SD of the Pierre Auger Observatory. The search for ES showers is performed considering the average signal Area over Peak over the triggered stations, which is related to the broadness of the shower front. As concerns DG showers, the neutrino selection is performed with a multivariate analysis by using different observables depending on the zenith angle, which are combined together in a Fisher discriminant. Assuming a differential flux $\phi = k \times E^{-2}$, the upper limit at 90% C.L. is set through the normalization k and is shown in Fig. 3. The integrated limit to k is 4.4×10^{-9} GeV cm^{-2} s^{-1}sr^{-1}, defined as the value needed for ~ 2.19 expected events, which is the Feldman-Cousins factor for non-observation of events in the absence of background. The maximum sensitivity of Auger is reached at ~ EeV, where most cosmogenic models peak, so that several models are already disfavoured by the current upper limits [12].

3.2 Neutrinos from point-like sources and transient events

Neutrinos from point-like sources can also be searched with the SD of Auger, with the same exposure calculation used for the diffuse flux but without integrating over the whole sky. A blind search for steady point-like sources is performed and upper limits as a function of declination are found for each data set, which are shown in Fig. 4. Being sensitive to energies above 10^{17} eV, Auger provides complementary information with respect to the limits given by other experiments, e.g. IceCube, which are sensitive to lower energies [13]. In order to look for neutrinos associated with transient events, the effective area over the relevant time interval has to be considered. The sensitivity to transient sources also depends on the efficiency of the detection during the time interval of interest and may exceed that of other dedicated neutrino detectors for some energies and some inclinations. The future detection of neutrinos in coincidence with a GW event could help in localizing the position of the source [14].

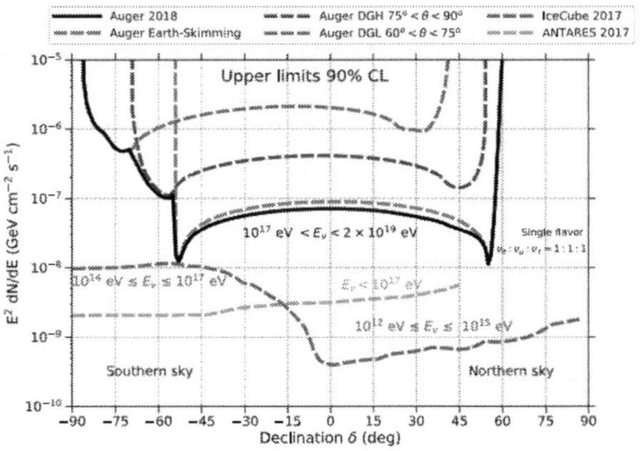

Figure 4. Upper limits at 90% C.L. to the normalization of the point-like flux of UHE neutrinos as a function of declination, together with results from other experiments [13].

4 Future prospects

The Pierre Auger Observatory will continue to monitor the UHE sky and contribute to multi-messenger studies. The on-going upgrade of the experiment, dubbed AugerPrime, will signficantly improve the discrimination between the electromagnetic and muonic components in air showers, which is a crucial aspect in the identification of neutral particles. Such an enhancement will further strengthen the capabilities of the Pierre Auger Observatory as a multi-messenger observatory in the near future [15].

References

[1] The Pierre Auger Collaboration, Nucl. Instrum. Methods Phys. Res. **798**, 172 (2015).
[2] The LIGO Collaboration and Virgo Collaboration, Phys. Rev. Lett. **119**, 161101 (2017).
[3] The IceCube Collaboration et al., Science **361**, eaat1378 (2018).
[4] M. Risse and P. Homola, Mod. Phys. Lett. **A22**, 749 (2007).
[5] The Pierre Auger Collaboration, Astrophys. J. **933**, 125 (2022).
[6] P. Savina for the Pierre Auger Collaboration, PoS (ICRC2021) **373** (2021).
[7] The Pierre Auger Collaboration, arXiv:2209.05926, JCAP (to be published).
[8] The Pierre Auger Collaboration, Universe **579**, 8(11) (2022).
[9] The Pierre Auger Collaboration, ApJ **789**, 160 (2014).
[10] The Pierre Auger Collaboration, ApJL **837**, L25 (2017).
[11] P. Ruehl for the Pierre Auger Collaboration, PoS (ICRC2021) **395** (2021).
[12] The Pierre Auger Collaboration, JCAP **10**, 022 (2019).
[13] The Pierre Auger Collaboration, JCAP **11**, 004 (2019).
[14] The Pierre Auger Collaboration, Phys. Rev. **D 94**, 122007 (2016).
[15] A. Castellina for the Pierre Auger Collaboration, EPJ Web Conf. **210** (2019).

Searching for high-energy neutrinos from the most luminous supernovae with the IceCube Neutrino Observatory

Massimiliano Lincetto[1,*] *for the IceCube Collaboration*

[1]Ruhr University Bochum, Faculty of Physics and Astronomy, Astronomical Institute (AIRUB), 44780 Bochum, Germany

Abstract.
The sources of the astrophysical neutrino flux discovered by IceCube remain for the most part unresolved. Extragalactic core-collapse supernovae (CCSNe) have been suggested as potentially able to produce high-energy neutrinos. In recent years, the Zwicky Transient Facility has discovered a population of exceptionally luminous supernovae, whose powering mechanisms have not yet been fully established. A fraction of these objects fall in the broader category of type IIn CCSNe, showing signs of interaction with a dense circumstellar medium. Theoretical models connect the supernova photometric properties to the dynamics of a shock-powered emission, predicting particle acceleration. In this contribution, we outline the plan for a search of high-energy neutrinos targeting the population of superluminous and type IIn supernovae with the IceCube Neutrino Observatory.

1 Introduction

The IceCube Neutrino Observatory is a large-volume neutrino detector based on an array of photomultipliers submerged in the Antarctic ice [1]. In the past decade, IceCube has discovered and characterised an astrophysical flux of high-energy neutrinos [2–4] which could be the key to identifying the origin of ultra-high-energy cosmic rays. While high-energy neutrino emission has been associated with different astrophysical objects, the majority of the flux observed by IceCube remains unresolved. Multi-messenger observations suggest that a significant fraction of the flux must originate in sources that are faint in high-energy gamma rays. Shock-powered transient phenomena observed in optical bands with suppressed gamma-ray emission are candidate neutrino sources [5]. Among these, type IIn core-collapse supernovae exhibit narrow lines in their spectrum that are a signature of interaction with a dense circumstellar medium. No high-energy neutrino excess has been previously observed from CCSNe with IceCube [6]. A search for high-energy neutrinos from a selected sample of supernovae powered by shock interaction could help in determining the origin of astrophysical neutrinos and in constraining the shock-interaction models for hadronic particle acceleration [7, 8].

2 The nature of the most luminous supernovae

In core-collapse supernovae (CCSNe), a large amount of gravitational energy is converted into low-energy neutrinos, thermal photons and kinetic energy of the ejecta. In Fig. 1, an

*e-mail: lincetto@astro.ruhr-uni-bochum.de

 https://doi.org/10.1051/epjconf/202328005005

overview of interesting subpopulations of CCSNe observed by the Zwicky Transient Facility Bright Transient Survey [9, 10] is given. An especially luminous subset of hydrogen-rich (type II) supernovae is represented by type IIn SNe, identified by narrow lines in their optical spectrum. This feature is a signal of interaction of the SN ejecta with the circumstellar medium (CSM). The shocks propagating in such environment may accelerate particles to relativistic energies. The interaction of the accelerated protons with the cold protons and nuclei in the dense CSM can result in the production of high-energy neutrinos [8] (see Fig. 2, left). Thanks to high-cadence wide-field astronomical surveys, populations of both hydrogen-poor (type I) and hydrogen-rich (type II) superluminous supernovae have been discovered in recent years [11]. Type II SLSNe are most likely interaction-powered and can be considered as an extension of the IIn population. Hydrogen-poor SLSNe represent a new, separate, class of SNe, likely powered by the combination of magnetar activity and CSM interaction. As both mechanisms can be responsible for particle acceleration, these sources will be considered in a future extension of this work. Type IIP SNe are excluded from this analysis, being less promising high-energy neutrino sources [7].

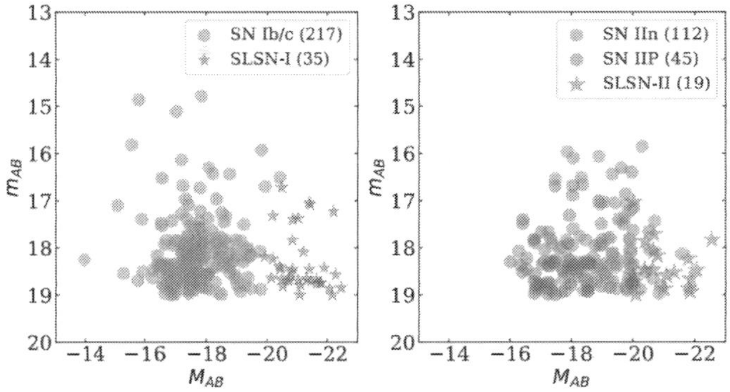

Figure 1. Distribution of the apparent magnitude (m_{AB}) as a function of absolute magnitude (M_{AB}) for selected subpopulations of type I (right) and type II (left) of core-collapse supernovae observed by the Zwicky Transient Facility in either ZTF-g or ZTF-r band. The following populations are omitted: Type-II SNe that lack a subtype classification, stripped-envelope SNe of type IIb, Ibn and Icn, and all subtypes classified as peculiar.

3 A sample of interaction-powered supernovae

A sample of type IIn SNe and type II SLSNe has been selected from the Bright Transient Survey of the Zwicky Transient Facility [9, 10]. The forced-photometry light curves for the ZTF red (ZTF-r) and green (ZTF-g) filters have been corrected for Galactic extinction and interpolated using Gaussian process regression. The sum of the interpolated light curves is taken as a *pseudo-bolometric* light curve, representing a lower limit on the optical emission of the source (see Fig. 2, right). We select events with robust spectroscopic classification, for which the time evolution has been well-observed in both bands, allowing for a reliable estimation of the rise and peak times of the supernova. The final selection includes 74 sources detected up to May 31, 2021 (end of the IceCube 2020-21 data season). The fluence at Earth for each source in the sample is calculated from the time-integrated pseudo-bolometric light curve and reported as a function of the source redshift and declination in Fig. 3.

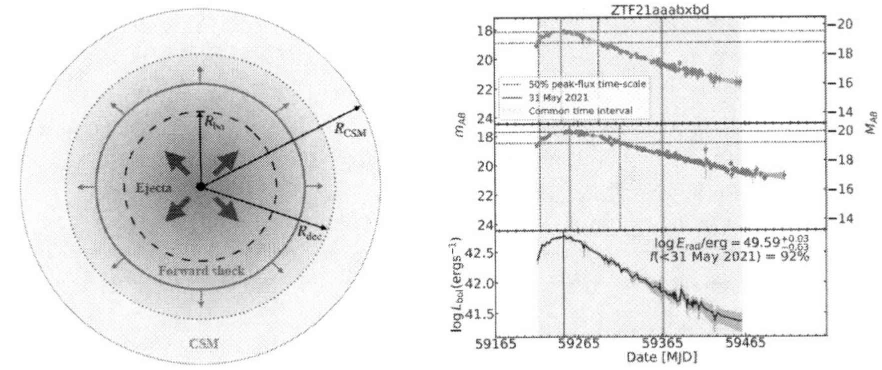

Figure 2. Left: sketch of the shock propagation in the circumstellar medium (CSM) for an interaction-powered supernova from Ref. [8]. Right: example of pseudo-bolometric light curve (bottom, see 3) obtained by the sum of interpolated ZTF-g (top) and ZTF-r (middle) band lightcurves of a type IIn CCSN observed by the Zwicky Transient Facility.

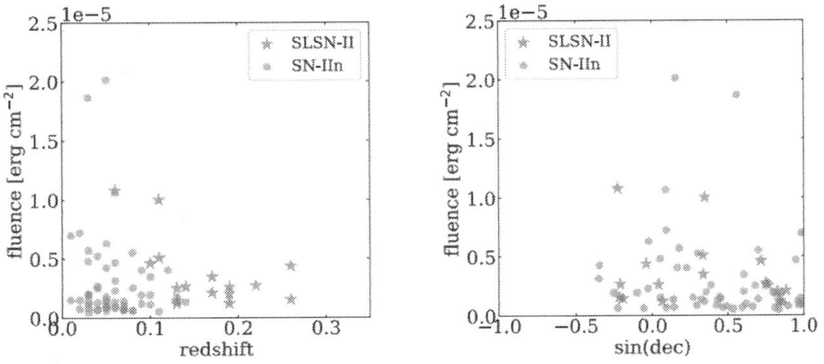

Figure 3. Pseudo-bolometric fluence as a function of redshift (left) and declination (right) for the selected sample of interaction-powered supernovae.

4 Neutrino stacking analysis and preliminary results

The analysis method for the search of astrophysical neutrinos is an unbinned stacking implemented in `flarestack` [12]. The likelihood function is defined on the neutrino sample:

$$\mathcal{L}(n_s, \gamma) = \prod_{i=0} \left[\frac{n_s}{N} \sum_{j=0}^{M} w_j \mathcal{S}_j(\theta_i, \gamma) + \left(1 - \frac{n_s}{N}\right) \mathcal{B}(\theta_i) \right] \tag{1}$$

where θ_i are the properties of the i-th neutrino (out of N), n_s is the total number of signal events from the source catalogue, w_j is the weight for source j (out of M), $\mathcal{S}$ and $\mathcal{B}$ are the signal and background probability density functions, γ is the spectral index of the signal flux. The likelihood $\mathcal{L}$ is maximised to find the best-fit values of n_s and γ. A test statistic is defined by the likelihood ratio of the *signal plus background* to the *background-only* hypotheses. The significance estimation is based on the test statistic distribution calculated from a large number of background neutrino samples, obtained by shuffling neutrino data in time and assigning random values of right ascension. Assuming a neutrino emission proportional to the optical luminosity, given that the two have a common powering mechanism, the sources are weighted according to their pseudo-bolometric fluence (see Sec. 3). A preliminary estimate of the sensitivity and discovery potential has been performed and is shown in Fig. 4. The corresponding neutrino energy fluxes are $4.5 \cdot 10^{-9}$ and $1.5 \cdot 10^{-8}$ $\mathrm{GeV^{-1}\ cm^{-2}\ s^{-1}}$. Improvements to the analysis are possible by restricting the catalogue to the Northern hemisphere (where IceCube has the best sensitivity) and by implementing a model-driven weighting of the sources based on a development of the approach described in Ref. [8].

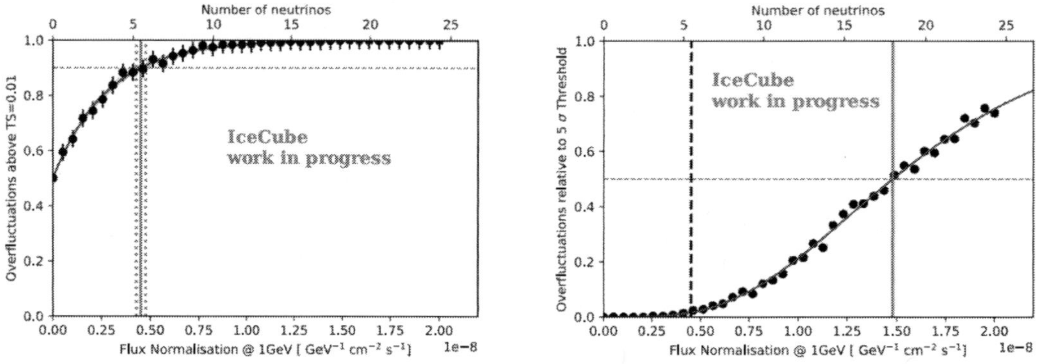

Figure 4. Sensitivity (flux at which 90% of the time the experiment yields a result above the background median; left) and discovery potential (flux at which 50% of the time the experiment yields a result above a $5\,\sigma$ significance; right) for the stacking analysis estimated by evaluating the test statistic distribution for different intensity of injected signal flux.

5 Acknowledgements

The authors thank Steve Schulze for the support provided in the analysis of the ZTF data and Tetyana Pitik for providing the theoretical framework that motivates this study. The authors acknowledge funding from the German Science Foundation DFG, within the Collaborative Research Center SFB-1491 "Cosmic Interacting Matters - From Source to Signal".

References

[1] M.G. Aartsen et al. (IceCube), JINST **12**, P03012 (2017), 1612.05093
[2] M.G. Aartsen et al. (IceCube), Astrophys. J. **833**, 3 (2016), 1607.08006
[3] M.G. Aartsen et al. (IceCube), Phys. Rev. Lett. **125**, 121104 (2020), 2001.09520
[4] R. Abbasi et al. (IceCube), Phys. Rev. D **104**, 022002 (2021), 2011.03545
[5] K. Fang, B.D. Metzger, I. Vurm, E. Aydi, L. Chomiuk, Astrophys. J. **904**, 4 (2020), 2007.15742
[6] J. Necker et al. (IceCube), PoS **ICRC2021**, 1116 (2021), 2107.09317
[7] K. Murase, Phys. Rev. D **97**, 081301 (2018), 1705.04750
[8] T. Pitik, I. Tamborra, C.R. Angus, K. Auchettl, Astrophys. J. **929**, 163 (2022), 2110.06944
[9] U.C. Fremling et al., Astrophys. J. **895**, 32 (2020), 1910.12973
[10] D.A. Perley et al., Astrophys. J. **904**, 35 (2020), 2009.01242
[11] A. Gal-Yam, Ann. Rev. Astron. Astrophys. **57**, 305 (2019), 1812.01428
[12] R. Stein, J. Necker, F. Bradascio, M. Lincetto, S. Garrappa, *icecube/flarestack: Titan v2.4.5* (2022), https://doi.org/10.5281/zenodo.6966491

Searching for Light Dark Matter with Aligned Carbon Nanotubes: the ANDROMeDa Project

Federico Virzi[1,4,*], *Alice* Apponi[3,6], *Gianluca* Cavoto[2,5], *Carlo* Mariani[2,5], *Francesco* Pandolfi[5], *Ilaria* Rago[2,5], *Alessandro* Ruocco[3,6]

[1]Università degli studi dell'Aquila
[2]Università degli studi di Roma La Sapienza
[3]Università degli studi di Roma 3
[4]INFN sezione di LNGS
[5]INFN sezione di Roma 1
[6]INFN sezione di Roma 3

Abstract. The ANDROMeDa (Aligned Nanotube Detector for Research On MeV Dark matter) project aims to develop a novel Dark Matter detector based on carbon nanotubes: the "Dark-PMT". The detector is designed to be sensitive to dark matter particles with mass between 1 MeV and 1 GeV. The detection scheme is based on dark matter-electron scattering inside a target made of vertically-aligned carbon nanotubes. Vertically-aligned carbon nanotubes have reduced density in the direction of the tube axis, therefore the scattered electrons are expected to leave the target without being re-absorbed only if their momentum has a small enough angle with that direction, which is what happens when the tubes are parallel to the dark matter wind. This grants directional sensitivity to the detector, a unique feature in this dark matter mass range.

1 Introduction

During the last decades several experiments have searched for dark matter (DM), typically for DM with mass $M_\chi > 10$ GeV [1]. In those experiments a high mass target has been used in order to have sensitivity for high mass DM particles. In fact, the detection scheme is based on measuring the interaction between the DM particle and the nucleus of the atom target. In a light DM scenario, instead, we have low sensitivity with current experiments because the momentum transferred to the nucleus from the DM-nucleus scattering is too small to be detected. In the low mass region the DM wind is expected to have kinetic energy T_χ in the range of $5 < T_\chi < 50$ eV for $10 < M_\chi < 100$ MeV, we also expect that DM particles come from the direction of the Cygnus constellation [2]. The ANDROMeDa project aims to measure DM in the low mass region by exploiting the important property of directionality of direct DM detection, this is done by using vertically-aligned Carbon NanoTubes (CNT) as target for DM-electron scattering. To do such measurement the project aims to develop a new detector: the "Dark-PMT".

*e-mail: μ federico.virzi@graduate.univaq.it

2 Carbon Nanotube

CNT can be described as graphene layer wrapped around an orientation axis with the property to be a 2D material with 1D properties and vanishing density along the axis. By exploiting this property the ANDROMeDa project aims to detect directly light DM, in fact the DM wind has a kinetic energy greater than the carbon work function ($\phi \simeq 4.7$ eV) and so it is possible to extract a recoil electron from the CNT lattice. As this is a non-relativistic scattering, the direction of the recoil electron will be, on average, parallel to the incoming DM wind. Vertically-aligned CNTs have vanishing density in the direction of the tubes and so the recoil electron is capable of exiting the target, without being re-absorbed, only if the tubes are aligned in the direction of the DM wind. This is shown in Figure 1 where it is depicted the distribution of the rates of recoil electrons that escape from the target at different kinetic energy for different θ_w [3], as can be noted the distributions for $\theta_w = 0°$ and $\theta_w = 90°$ have different rates and so it is possible to distinguish them.

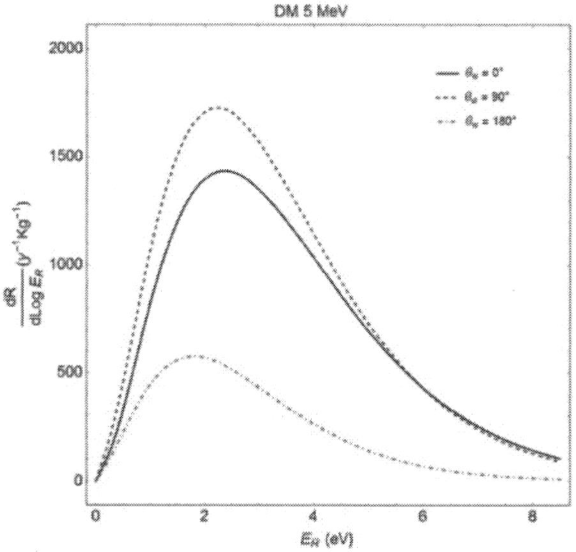

Figure 1. Electron energy recoil spectra (per year per kilogram target) for $M_\chi = 5$ MeV. Three relative orientations θ_w between the nanotube axes and the DM wind are shown: $\theta_w = 0°$ (in black, solid), $\theta_w = 90°$ (in blue, dashed) and $\theta_w = 180°$ (in red, dot-dashed) [3].

With this experimental idea it is important to produce well vertically aligned CNT, otherwise electron re-absorption can occur. CNT can be grown in forest of the surface of O(8 cm^2) with Chemical Vapor Deposition (CVD) technique in the facility that has been assembled and installed by INFN in 2020 in Rome la Sapienza, the CNT sample has a length of almost 260 μm as can be seen in Figure 2. During the growing process of the CNT sample two undesired effects can happen: the waviness modulation of the CNT and the formation of a crust layer at the top of the CNT forest. Considering the low kinetic energy of the recoil electron, the two effects described above increase the possibility of electron re-absorption. This concept of CNT target has been tested by measuring the degradation due to positive ions bombardment for transverse and longitudinal direction with respect to the CNT orientation axis [4]. The most important R&D activity of the ANDROMeDa project is to grow vertically aligned CNT samples minimizing both the undesired effects.

A possibility to minimize the waviness of the CNT sample is to optimize the growing process by increasing the seeds density with uniform seeds and to grow the CNT in an electric field, the crust layer, instead, can be removed by using a plasma etching [5].

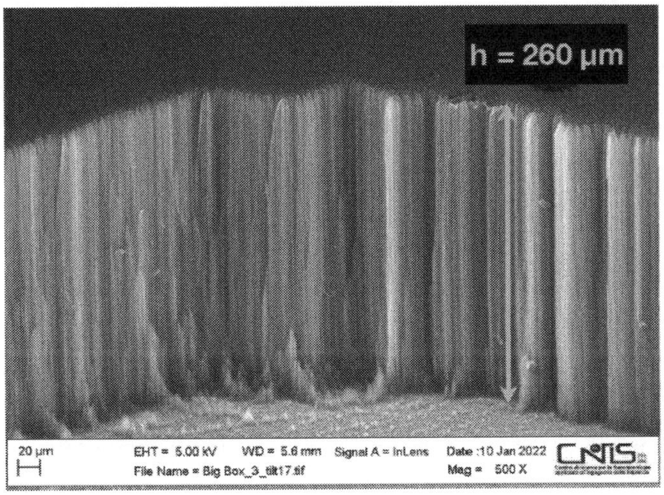

Figure 2. Typical growth of vertically aligned CNT forest, as imaged by Scanning Electron Microscopy.

3 ANDROMeDa Dark-PMT

The usage of CNT as target for light DM detection can be exploited by developing a "Dark-PMT" detector. The scheme of the Dark-PMT prototype is shown in Figure 3. When the DM wind interacts with the target produces a recoil electron, this electron escapes from the target if the DM wind is aligned with the CNT target. In this case the electron has a kinetic energy of almost 5 eV and it is then accelerated to 5 keV by using an electric field. The electron's energy is measured by using silicon avalanche photodiode (APD) that are adjusted for electron usage. APD has been characterized for electron detection by using a monoenergetic electron gun of energy $[90, 900]$ eV in the LASEC laboratory of Roma Tre university [6]. In the Dark-PMT setup it is important to being able to distinguish each single electron from the target, in this way it is possible to reject events from cosmic rays. In fact a cosmic ray event could generate various ionization electrons in the target while a DM event just one electron. The good point of this setup is that the device grant directional sensitivity, is cheap and easy to produce, has compact size and can be used at room temperature because it is not affected by thermal noise. This detector can be used for a future experiment by assembling various arrays of two Dark-PMTs, one towards the opposite direction of the other. By pointing the array towards Cygnus, it is possible to measure the rates of electrons that exit from the target as is shown in Figure 1 while the PMTs oriented in the opposite direction measure the background. Considering a CNT density of ρ = 1 mg/cm^2 with 10 cm^2 surface, with 100 Dark-PMTs ANDROMeDa aims to reach the expected sensitivity for 1 g× yr exposure that is shown in Figure 4. As can be noted with just 1 g target (red line) the experiment is competitive with XENON exclusion limits [7] for $M_\chi \sim 40$ MeV while with future upgrade with a target mass of 1 Kg (blue line) the exclusion limits are below for the entire mass range that is shown in Figure 4 .

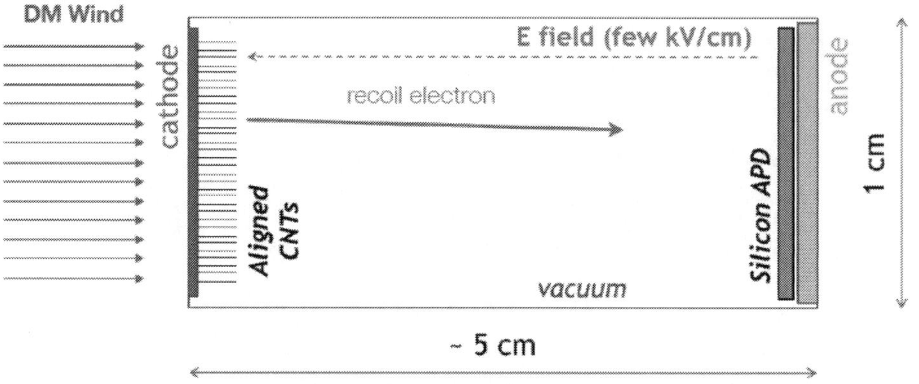

Figure 3. ANDROMeDa scheme of a "Dark-PMT". The DM wind interact with the CNT target and then the recoil electron is accelerated to 5 keV and is detected by using silicon APD.

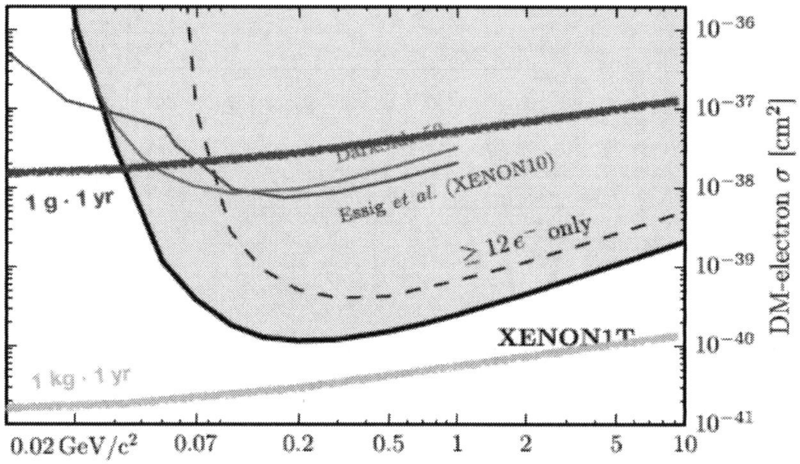

Figure 4. ANDROMeDa expected exclusion limits (red and blu lines). On the x axis is shown the DM mass while on the y axis is shown the DM-electron cross section. Figure adapted from [7].

References

[1] E.Aprile et al., Physical Review D **100**, 052014 (2019)
[2] Ciaran A.J. O'Hare et al., Physical Review D **92**, 063518 (2015)
[3] G.Cavoto, F. Luchetta, A.D.Polosa, Physics Letters B **776**, 338-344 (2018)
[4] G.D'Acunto et al., Carbon **139**, 768-775 (2018)
[5] S.Seo et al., Carbon **180**, 204-214 (2021)
[6] A.Apponi et al., Instrumentation and Detectors **arXiv:2008.07169v2**, (2020)
[7] E.Aprile et al., Physical Review letters **123**, 251801 (2019)
[8] A.Apponi et al., JINST **15**, P11015 (2020)
[9] L.M.Capparelli et al., Physics of the Dark Universe **9-10**, 24-30 (2015)

Light dark matter detection: new ideas and new tools

Angelo Esposito[1,2,*]

[1]Dipartimento di Fisica, Sapienza Università di Roma, Piazzale Aldo Moro 2, I-00185 Rome, Italy
[2]INFN Sezione di Roma, Piazzale Aldo Moro 2, I-00185 Rome, Italy

Abstract. We discuss possibile ways to probe dark matter in the keV to GeV range with direct detection experiments.

1 Introduction

There is today overwhelming evidence that the majority of matter in the Universe is dark matter, feebly interacting with ordinary matter. Despite the crucial role it plays in our understanding of the Universe, we still know very little about its nature. In particular, its possible mass ranges over more that sixty orders of magnitude. Following stringent constraints on heavy WIMPs [e.g., 1–5], there is now increasing interest in models for sub-GeV dark matter [e.g., 6–8], motivating new detection ideas. In particular, dark matter candidates in the keV to GeV range cannot release appreciable energy via standard nuclear recoil, and thus require detectors with low energy thresholds, such as semiconductors [e.g., 10–12], superconductors [e.g., 13–15], lower dimensional materials [e.g., 16, 17], and so on [e.g., 18–20].

Here we focus on some possible ways to push the sensitivity of direct detection experiments down to masses of order keV. Specifically, for the range $1\text{ MeV} \lesssim m_\chi \lesssim 1\text{ GeV}$, we discuss the so-called Migdal effect in semiconductors [21–24]. For the range $1\text{ keV} \lesssim m_\chi \lesssim 1\text{ MeV}$ we, instead, discuss signatures involving phonons in superfluid ^{4}He [e.g., 25–32] or magnons in anti-ferromagnetic materials [33]. We describe the processes above using effective field theory (EFT) methods, which allow to bypass some difficulties encountered in the more traditional approaches, and simply the theoretical treatment.

2 MeV to GeV: Migdal effect in semiconductors

Semiconductors, like Si or Ge, have bandgaps of the order of a few eV, making them promising targets to hunt for dark matter particles as light as an MeV. When the dark matter couples mostly to the nuclei, it can promote an electron from the valence band to the conduction one via the Migdal effect: the dark matter interacts with a nucleus, which in turns interacts with the electrons through lattice excitations. Computing the rate for this event in a realistic material is complicated by the many-body, strongly interacting dynamics of both nuclei and electrons. So far, the problem has been treated under some simplifying approximations, namely the impulse [21] and harmonic [22, 23] ones. The former is reliable for $m_\chi \gtrsim 50$ MeV, while the latter for $m_\chi \simeq 1$ MeV. It is necessary to develop a formalism for the general description of the phenomenon, which allows for sound predictions for all dark matter masses.

*e-mail: angelo.esposito@uniroma1.it

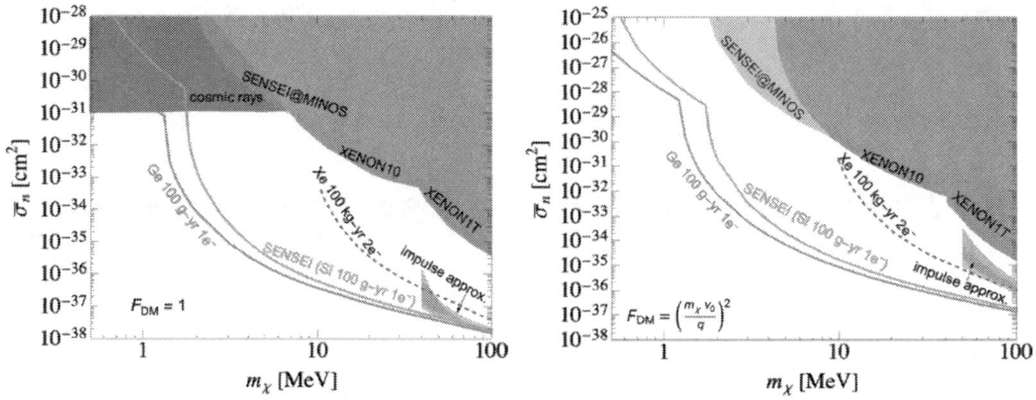

Figure 1. Projections for the dark matter–nucleon cross section using the Migdal effect at 90% C.L., assuming an exposure of 100-g-year of both Si (**cyan**) and Ge (**red**). In both cases, we assume a detector sensitive to single electrons excitations, as well as zero background. For details see [24].

As noticed in [24], the problem at hand is charachterized by a separation of scales: the energy of the Migdal electron is O(eV), much larger than the typical energy of the excitations of the crystal, which is $O(1-100$ meV). This allows to bypass the complicated description of the crystal dynamics by writing a low energy EFT, which describes the Migdal emission by the dark matter via the following effective Hamiltonian [24]:

$$H_{\rm eff} = \frac{1}{m_{\rm N}\omega^2} \nabla H_{\chi {\rm L}} \cdot \nabla H_{e{\rm L}} + O\left(1/\omega^3\right) , \tag{1}$$

where ω is the energy of the outgoing Migdal electron, $H_{\chi\rm L}$ the Hamiltonian coupling the dark matter to the nuclei in the crystal, and $H_{e\rm L}$ the one coupling the nuclei to the electrons. Thanks to this, it is possible to show that the rate for Migdal emission, differential both in the electron energy, ω, as well as in the energy released to the crystal, $E_{\rm ph}$, is [24],

$$\frac{d^2\Gamma}{d\omega dE_{\rm ph}} \propto \sum_{\boldsymbol{k}} \sum_{\boldsymbol{K},\boldsymbol{Q}} \frac{\boldsymbol{q}\cdot(\boldsymbol{k}+\boldsymbol{K})\, \boldsymbol{q}\cdot(\boldsymbol{k}+\boldsymbol{K})}{|\boldsymbol{k}+\boldsymbol{K}||\boldsymbol{k}+\boldsymbol{Q}|} {\rm Im}\left(-\epsilon^{-1}_{\boldsymbol{K}\boldsymbol{Q}}(\boldsymbol{k},\omega)\right) S(\boldsymbol{q}-\boldsymbol{k}-\boldsymbol{K}, E_{\rm ph}) . \tag{2}$$

Here $\boldsymbol{q}$ is the momentum transfer by the dark matter, $\boldsymbol{K}$ and $\boldsymbol{Q}$ are momenta belonging to the reciprocal lattice, and $\boldsymbol{k}$ is the crystal momentum of the outgoing electron (belonging to the first Brillouin zone). The electron response of the material is encoded in the energy loss function (ELF) [e.g., 34–36], ${\rm Im}\left(-\epsilon^{-1}_{\boldsymbol{K}\boldsymbol{Q}}(\boldsymbol{k},\omega)\right)$, while the crystal response in the so-called structure factor [e.g., 37], $S(\boldsymbol{q}, E)$. Both these quantities are, a priori, measurable from data, thus bypassing the need for an accurate modelling of the nuclei and electrons in the semiconductor.

With this result at hand, one can compute the projected limits obtained from the Migdal effect in case of SENSEI [e.g., 38], which employs Si, as well as of a possible detector based on Ge. The results are reported in Figure 1. For measurements which are inclusive on the energy released to the crystal, the rate is insensitive to the details of the crystal under consideration [24]. Possible future measurements, sensitive also to $E_{\rm ph}$, could instead be used as background discrimination, making it imperative to have reliable predictions based on Eq. (2), and, consequently, to precisely measured both the ELF and the structure factor.

3 keV to MeV: superfluid ^{4}He and anti-ferromagnets

When looking for dark matter with masses $m_\chi \lesssim O(\mathrm{MeV})$, the typical momentum exchange is $1/q \gtrsim O(\text{Å})$, larger than the interparticle separation in the target material. Consequently, the interaction with dark matter likely produces collective excitations, rather than single particle ones. Depending on the type of interaction of dark matter with the Standard Model, it can couple more efficiently to different kinds of collective excitations. Here we discuss the possibility of probing spin-independent interactions using phonons in superfluid ^{4}He [28, 29, 31, 39], as well as spin-dependent one using magnons in anti-ferromagnetic materials [33]. Our theoretical description relies on the fact that these collective modes, at low energies, are Goldstone bosons associated to some spontaneously broken symmetries. As such, their interactions can be solely described in terms of symmetries and a handful of effective parameters.

3.1 Phonons in superfluid ^{4}He

The ground state of a superfluid (like ^{4}He) spontaneously breaks boosts, time translations and the particle number symmetry, preserving a linear combination of the last two [e.g., 40]. The associated Goldstone is the phonon, $\pi(x)$, corresponding to the material density perturbations. At low energies, its self-interactions are described by an effective Lagrangian [e.g., 39],

$$\mathcal{L}_\pi = \frac{1}{2}\dot{\pi}^2 - \frac{c_s^2}{2}(\nabla\pi)^2 + \frac{\lambda}{6}\dot{\pi}^3 + \frac{\lambda'}{2}\dot{\pi}(\nabla\pi)^2 + O\left(\pi^4\right), \tag{3}$$

where c_s is the speed of sound in the superfluid, and λ and λ' are effective coefficients that can be determined from the equation of state [28, 29, 31]. As far as the dark matter–phonon interaction is concerned, many natural dark matter models lead, in the non-relativistic limit, to a simple coupling between the dark matter and the superfluid density, n, [e.g., 28, 31]:

$$\mathcal{L}_{\rm int} \propto \chi^\dagger \chi n(\pi) \propto \chi^\dagger \chi \left[\alpha\dot{\pi} + \frac{\beta_1}{2}(\nabla\pi)^2 + \frac{\beta_2}{2}\dot{\pi}^2 + O\left(\pi^3\right)\right], \tag{4}$$

with χ the non-relativistic dark matter field, and α and $\beta_{1,2}$ given by the equation of state [31].

With this at hand, the event rate for the emission of any number of low energy phonons can be computed with standard diagramatic methods. This has been used to determine the rate of emission of two phonons, as first suggested in [26, 27], which allows to find the projected exclusion plots for masses down to $m_\chi \sim O(\mathrm{keV})$ [28, 29]. A similar analysis has also been performed for a model of dark matter which interacts with the Standard Model via a new Abelian vector mediator [39]. Finally, it has been used to compute the rate of a more involved process: the emission of three phonons [31]. While the latter is expected to be suppressed with respect to a two-phonon event, it offers an intriguing possibility for background discrimination. Indeed, it allows for a configuration with two back-to-back phonons, which can be used for coincidence, and one phonon emitted in the same direction as the incoming dark matter, which can instead provide directional information. The previous EFT can be used to determine the angular distributions of such an event, as well as its total rate, showing that it actually dominates over the two-phonon event in the $0.7\ \mathrm{MeV} \lesssim m_\chi \lesssim 1\ \mathrm{MeV}$ range [31].

3.2 Magnons in anti-ferromagnets

If dark matter, instead, features dominant spin-dependent interaction, a more suitable class of target materials is anti-ferromagnets:[1] the dark matter couples to the individual spins in the

[1]Ferromagnets have also been considered [41–44]. Anti-ferromagnets, however, can probe masses an order of magnitude lighter [33].

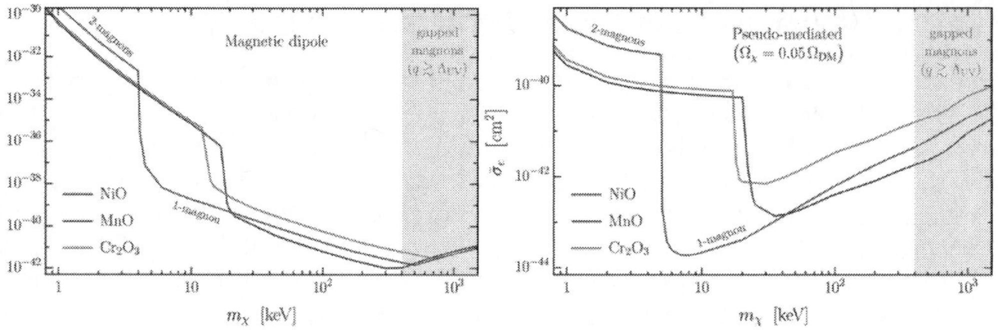

Figure 2. Projected exclusion limits for the dark matter–electron cross section at 90% C.L. for three different anti-ferromagnets and two benchmark dark matter models, assuming zero background. See [33] for details.

material, causing the emission of spin waves, called magnons. When gapless, they too are Goldstones, following from the fact that the ground state of an anti-ferromagnet (all spins are anti-aligned in one direction) spontaneously breaks the internal $SO(3)$ spin symmetry, down to an unbroken $SO(2)$. Similarly to the case of phonons in a superfluid, low energy magnons in an anti-ferromagnet can be described by the following effective Lagrangian [45, 46],

$$\mathcal{L}_\theta = \frac{c_1}{2}\left(\dot{\theta}^a\right)^2 - \frac{c_2}{2}\left(\partial_i\theta^a\right)^2 + O\left(\theta^3\right), \tag{5}$$

with $\theta^a(x)$ the magnon field ($a = 1, 2$), and $c_{1,2}$ effective coefficients which are determined from the magnon speed of propagation together with neutron scattering data [33]. The coupling between dark matter and magnons, instead, happens via the spin density operator, $s(x)$:

$$\mathcal{L}_{\text{int}} \propto \left(\chi^\dagger\hat{\Gamma}^i\chi\right)\left(\hat{\mathcal{D}}_{ij}s^j(\theta)\right) = c_1\left(\chi^\dagger\hat{\Gamma}^i\chi\right)\left[\hat{\mathcal{D}}_{ij}\left(\delta^j_a\theta^a + \delta^j_3\epsilon_{ab}\theta^a\dot{\theta}^b + O\left(\theta^3\right)\right)\right], \tag{6}$$

where $\hat{\Gamma}$ is a possible spin operator and $\hat{\mathcal{D}}$ a possible differential operator. Both of them depend on the particular dark matter model under consideration.

Again, we can now use textbook techniques to evaluate the event rate for the emission of any number of magnons. In Figure 2 we report the projected exclusion plots fro three different anti-ferromagnets and two benchmark dark matter models. Firstly, all anti-ferromagnets allows to probe masses as light as the keV. Secondly, one of them, NiO, feature a speed of propagation surprisingly close to the typical dark matter velocity, making it particularly efficient at absorbing most of its energy, already via the dominant one-magnon emission process.

4 Conclusion

The hunt for dark matter in the keV to GeV range requires new detectors ideas and, consequently, new theoretical tools for the description of the relevant phenomena. Taking advantage of the Migdal effect in semiconductors, one can extend the sensitivity to the dark matter–nucleon interaction down to masses $m_\chi \sim O(\text{MeV})$. Signatures involving collective modes in material such as superfluid ^{4}He and anti-ferromagnets, instead, can push the experimental sensitivity down to $m_\chi \sim O(\text{keV})$. In all these instances, one must deal with the richness and complicacy of condensed matter phenomena. One possible way to account for that is using effective field theory methods, as discussed here.

References

[1] R. Agnese *et al.* [SuperCDMS], Phys. Rev. Lett. **116**, no.7, 071301 (2016) [arXiv:1509.02448 [astro-ph.CO]].

[2] D. S. Akerib *et al.* [LUX], Phys. Rev. Lett. **118**, no.2, 021303 (2017) [arXiv:1608.07648 [astro-ph.CO]].

[3] E. Aprile *et al.* [XENON], Phys. Rev. Lett. **123**, no.25, 251801 (2019) [arXiv:1907.11485 [hep-ex]].

[4] A. H. Abdelhameed *et al.* [CRESST], Phys. Rev. D **100**, no.10, 102002 (2019) [arXiv:1904.00498 [astro-ph.CO]].

[5] Q. Wang *et al.* [PandaX-II], Chin. Phys. C **44**, no.12, 125001 (2020) doi:10.1088/1674-1137/abb658 [arXiv:2007.15469 [astro-ph.CO]].

[6] C. Boehm, P. Fayet and J. Silk, Phys. Rev. D **69**, 101302 (2004) [arXiv:hep-ph/0311143 [hep-ph]].

[7] M. J. Strassler and K. M. Zurek, Phys. Lett. B **651**, 374-379 (2007) [arXiv:hep-ph/0604261 [hep-ph]].

[8] J. L. Feng and J. Kumar, Phys. Rev. Lett. **101**, 231301 (2008) [arXiv:0803.4196 [hep-ph]].

[9] R. Essig, J. Mardon and T. Volansky, Phys. Rev. D **85**, 076007 (2012) [arXiv:1108.5383 [hep-ph]].

[10] P. W. Graham, D. E. Kaplan, S. Rajendran and M. T. Walters, Phys. Dark Univ. **1**, 32-49 (2012) [arXiv:1203.2531 [hep-ph]].

[11] R. Essig, M. Fernandez-Serra, J. Mardon, A. Soto, T. Volansky and T. T. Yu, JHEP **05**, 046 (2016) [arXiv:1509.01598 [hep-ph]].

[12] Y. Hochberg, T. Lin and K. M. Zurek, Phys. Rev. D **95**, no.2, 023013 (2017) [arXiv:1608.01994 [hep-ph]].

[13] Y. Hochberg, Y. Zhao and K. M. Zurek, Phys. Rev. Lett. **116**, no.1, 011301 (2016) [arXiv:1504.07237 [hep-ph]].

[14] Y. Hochberg, I. Charaev, S. W. Nam, V. Verma, M. Colangelo and K. K. Berggren, Phys. Rev. Lett. **123**, no.15, 151802 (2019) [arXiv:1903.05101 [hep-ph]].

[15] S. M. Griffin, Y. Hochberg, K. Inzani, N. Kurinsky, T. Lin and T. Chin, Phys. Rev. D **103**, no.7, 075002 (2021) [arXiv:2008.08560 [hep-ph]].

[16] G. Cavoto, F. Luchetta and A. D. Polosa, Phys. Lett. B **776**, 338-344 (2018) [arXiv:1706.02487 [hep-ph]].

[17] Y. Hochberg, Y. Kahn, M. Lisanti, C. G. Tully and K. M. Zurek, Phys. Lett. B **772**, 239-246 (2017) [arXiv:1606.08849 [hep-ph]].

[18] S. Knapen, T. Lin, M. Pyle and K. M. Zurek, Phys. Lett. B **785**, 386-390 (2018) [arXiv:1712.06598 [hep-ph]].

[19] B. Campbell-Deem, P. Cox, S. Knapen, T. Lin and T. Melia, Phys. Rev. D **101**, no.3, 036006 (2020) [erratum: Phys. Rev. D **102**, no.1, 019904 (2020)] [arXiv:1911.03482 [hep-ph]].

[20] B. Campbell-Deem, S. Knapen, T. Lin and E. Villarama, Phys. Rev. D **106**, no.3, 036019 (2022) [arXiv:2205.02250 [hep-ph]].

[21] S. Knapen, J. Kozaczuk and T. Lin, Phys. Rev. Lett. **127**, no.8, 081805 (2021) [arXiv:2011.09496 [hep-ph]].

[22] Z. L. Liang, C. Mo, F. Zheng and P. Zhang, Phys. Rev. D **106**, no.4, 043004 (2022) [erratum: Phys. Rev. D **106**, no.10, 109901 (2022)] [arXiv:2205.03395 [hep-ph]].

[23] Z. L. Liang, C. Mo, F. Zheng and P. Zhang, Phys. Rev. D **104**, no.5, 056009 (2021) [arXiv:2011.13352 [hep-ph]].

[24] K. V. Berghaus, A. Esposito, R. Essig and M. Sholapurkar, [arXiv:2210.06490 [hep-ph]].

[25] W. Guo and D. N. McKinsey, Phys. Rev. D **87**, no.11, 115001 (2013) [arXiv:1302.0534 [astro-ph.IM]].

[26] K. Schutz and K. M. Zurek, Phys. Rev. Lett. **117**, no.12, 121302 (2016) [arXiv:1604.08206 [hep-ph]].

[27] S. Knapen, T. Lin and K. M. Zurek, Phys. Rev. D **95**, no.5, 056019 (2017) [arXiv:1611.06228 [hep-ph]].

[28] F. Acanfora, A. Esposito and A. D. Polosa, Eur. Phys. J. C **79**, no.7, 549 (2019) [arXiv:1902.02361 [hep-ph]].

[29] A. Caputo, A. Esposito and A. D. Polosa, Phys. Rev. D **100**, no.11, 116007 (2019) [arXiv:1907.10635 [hep-ph]].

[30] G. Baym, D. H. Beck, J. P. Filippini, C. J. Pethick and J. Shelton, Phys. Rev. D **102**, no.3, 035014 (2020) [erratum: Phys. Rev. D **104**, no.1, 019901 (2021)] [arXiv:2005.08824 [hep-ph]].

[31] A. Caputo, A. Esposito, F. Piccinini, A. D. Polosa and G. Rossi, Phys. Rev. D **103**, no.5, 055017 (2021) [arXiv:2012.01432 [hep-ph]].

[32] B. von Krosigk, K. Eitel, C. Enss, T. Ferber, L. Gastaldo, F. Kahlhoefer, S. Kempf, M. Klute, S. Lindemann and M. Schumann, *et al.* [arXiv:2209.10950 [hep-ex]].

[33] A. Esposito and S. Pavaskar, [arXiv:2210.13516 [hep-ph]].

[34] S. Knapen, J. Kozaczuk and T. Lin, Phys. Rev. D **104**, no.1, 015031 (2021) [arXiv:2101.08275 [hep-ph]].

[35] Y. Hochberg, Y. Kahn, N. Kurinsky, B. V. Lehmann, T. C. Yu and K. K. Berggren, Phys. Rev. Lett. **127**, no.15, 151802 (2021) [arXiv:2101.08263 [hep-ph]].

[36] S. Knapen, J. Kozaczuk and T. Lin, Phys. Rev. D **105**, no.1, 015014 (2022) [arXiv:2104.12786 [hep-ph]].

[37] G. L. Squres, "Introduction to the Theory of Thermal Neutron Scattering," Cambridge University Press.

[38] L. Barak *et al.* [SENSEI], Phys. Rev. Lett. **125**, no.17, 171802 (2020) [arXiv:2004.11378 [astro-ph.CO]].

[39] A. Caputo, A. Esposito, E. Geoffray, A. D. Polosa and S. Sun, Phys. Lett. B **802**, 135258 (2020) [arXiv:1911.04511 [hep-ph]].

[40] A. Nicolis, R. Penco, F. Piazza and R. Rattazzi, JHEP **06**, 155 (2015) [arXiv:1501.03845 [hep-th]].

[41] T. Trickle, Z. Zhang and K. M. Zurek, Phys. Rev. Lett. **124**, no.20, 201801 (2020) [arXiv:1905.13744 [hep-ph]].

[42] A. Mitridate, T. Trickle, Z. Zhang and K. M. Zurek, Phys. Rev. D **102**, no.9, 095005 (2020) [arXiv:2005.10256 [hep-ph]].

[43] D. J. E. Marsh, K. C. Fong, E. W. Lentz, L. Smejkal and M. N. Ali, Phys. Rev. Lett. **123**, no.12, 121601 (2019) [arXiv:1807.08810 [hep-ph]].

[44] S. Chigusa, T. Moroi and K. Nakayama, Phys. Rev. D **101**, no.9, 096013 (2020) doi:10.1103/PhysRevD.101.096013 [arXiv:2001.10666 [hep-ph]].

[45] C. P. Burgess, Phys. Rept. **330**, 193-261 (2000) [arXiv:hep-th/9808176 [hep-th]].

[46] S. Pavaskar, R. Penco and I. Z. Rothstein, SciPost Phys. **12**, no.5, 155 (2022) [arXiv:2112.13873 [hep-th]].

Direct Detection of Dark Matter with DarkSide-20k

Paolo Agnes[1,*], on behalf of the *DarkSide Collaboration*

[1]Gran Sasso Science Institute, viale Francesco Crispi 7, 67100, L'Aquila, IT

Abstract. DarkSide run since mid 2015 a 50-kg-active-mass dual phase Liquid Argon Time Projection Chamber (TPC), filled with low radioactivity argon from an underground source and produced world class results for both the low mass (< 20 GeV/c^2) and high mass (> 100 GeV/c^2) direct detection search for dark matter. The next stage of the DarkSide program will be a new generation experiment involving a global collaboration from all the current Argon based experiments. DarkSide-20k, is designed as a 20-tonne fiducial mass dual phase Liquid Argon TPC with SiPM based cryogenic photosensors, and is expected to be free of any instrumental background for an exposure of >100 tonne year. Like its predecessor, DarkSide-20k will be housed at the INFN Gran Sasso (LNGS) underground laboratory, and it is expected to attain a WIMP-nucleon cross section exclusion sensitivity of 7.4×10^{-48} cm^2 for a WIMP mass of 1 TeV/c^2 in a 200 t yr run. DarkSide-20k will be installed inside a membrane cryostat containing more than 700 t of liquid Argon and be surrounded by an active neutron veto based on a Gd-loaded acrylic shell. The talk will give the latest updates of the ongoing R&D and prototype tests validating the initial design. A subsequent objective, towards the end of the next decade, will be the construction of the ultimate detector, ARGO, with a 300 t fiducial mass to push the sensitivity to the neutrino floor region for high mass WIMPs.

1 Direct Dark Matter Search with Liquid Argon

There is strong evidence from astronomical and cosmological observations for the existence of dark matter in our Universe. Weakly Interacting Massive Particles (WIMPs) are a well-motivated dark matter candidate that may have been produced in the early Universe and would show the correct relic abundance observed today. They are however so weakly interacting that they have yet to be observed in a terrestrial experiment. Earth-based direct detection experiments aim at detect galactic dark matter particles via their interaction with ordinary matter in an instrumented target. Noble liquid detectors, in the dual phase Time Projection Chamber (TPC) layout, are particularly suitable for these searches, thanks to scalability, intrinsic radio-purity and good ionization and scintillation properties of the target. The TPC consists of an active volume filled with liquid, which is immersed in a uniform electric field, to drift ionization electrons towards a gaseous region on top of the detector, where charges are accelerated to stimulate light production by electro-luminescence. Photo-detectors collect both the prompt scintillation light produced in liquid (S1), and the delayed electro-luminescence light (S2) generated in gas, allowing for energy measurement and 3D vertex reconstruction.

*e-mail: paolo.agnes@gssi.it

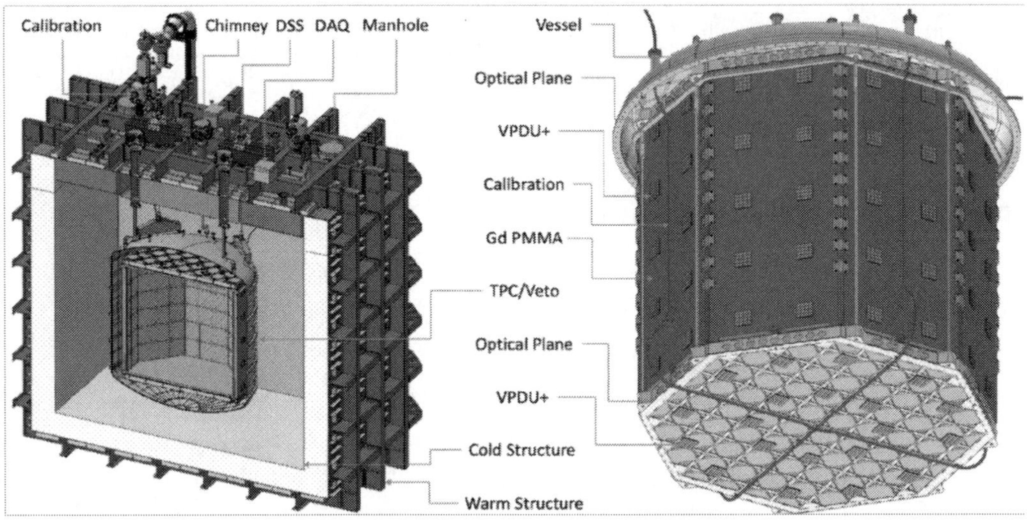

Figure 1. Schematics of the whole DarkSide-20k experimental setup to be installed in Hall C at LNGS (left). A closeup of the inner detector is shown on the right.

In addition to this, liquid argon (LAr) has excellent event discrimination capabilities: scintillation light initiated by particles recoiling from atomic electrons, that is the primary source of background in a WIMP direct detection experiment, has a time constant of approximately 1 μs. This is in stark contrast to the nanosecond time constant of scintillation light emitted during an expected WIMP-nuclear recoil event. The DEAP-3600 experiment has exploited this effect via pulse shape discrimination (PSD) to achieve electron recoil background rejection of 2.4×10^8 [1]. DarkSide-50, whose experimental layout is described in [2], has presented results from a 532,4 live-days exposure, and observed no background events over a run period in excess of two years [3]. In addition to sensitivity to WIMPs with masses above 100 GeV/c^2, the two-phase DarkSide-50 detector has extended its reach to WIMP masses below 10 GeV/c^2 by detecting single ionizaton electrons extracted from the liquid argon volume [4–6]. These results were possible as argon extracted from deep underground sources (UAr) was used as the active medium, to avoid the $\sim$ 1 Bq/kg ^{39}Ar activity in atmospheric argon produced by cosmic-ray activation. DarkSide-50 has demonstrated that using underground argon the rate of ^{39}Ar events could be reduced by a factor of 1400 ± 200 [7].

2 The DarkSide-20k detector

The DarkSide-20k detector consists of three nested detectors housed within a ProtoDUNE-style membrane cryostat [8] as shown in the sketch in Figure 1. The inner detector is a dual-phase argon TPC, filled with 49.of UAr. The TPC is inserted in a stainless steel vessel, filled with an additional 36 t of UAr, acting as radiogenic neutron veto. The AAr filling the cryostat (700 t) is instrumented to detect scintillation light from cosmic muons crossing the volume.

2.1 The procurement of Underground Argon

The journey of the DarkSide-20k UAr starts in Cortez, CO. It is extracted from underground as a byproduct of CO_2 in an industrial extraction plant at a rate of 250-330 kg/day, in the

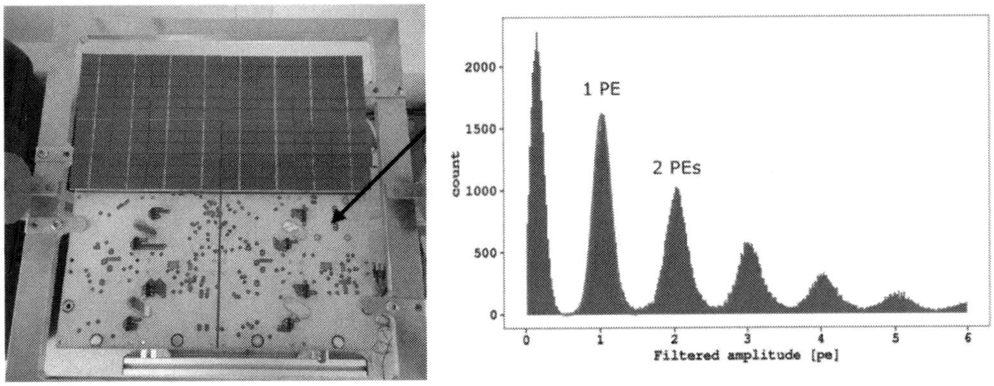

Figure 2. Left: an assembly of 4 PDMs being populated. Right: a typical charge spectrum from a single PDMs, showing the excellent singe photoelectron resolution.

URANIA facility. The expected purity at the output is 99.99%. Once extracted, the UAr will travel to Europe, mainly at the surface level to minimize the s activation during transport. The UAr will reach the Carbosulcis mine, in Sardinia, where the 350 m tall ARIA distillation column will be operated. ARIA will be able to further chemically purify the UAr and removing rare stable isotopes at a rate of 1 t/day [9]. Beyond DarkSide-20k, ARIA has the potential to further reduce the ^{39}Ar content in the target by a factor of 10 by cryogenic distillation at the slower rate of 10 kg/day. A sample of the UAr production will be shipped to the Canfranc Underground Laboratory, where the DArT detector, installed inside ArDM, will be used to characterize the residual ^{39}Ar with target uncertainty of 10 % in few weeks of data taking [10].

2.2 The DarkSide-20k TPC

The active volume of the TPC is defined by eight vertical reflector panels (1.5 m wide, 3.5 tall) and the top and bottom windows of the acrylic vessel. The top and bottom caps of the vessel are made of ultra-pure acrylic (PMMA), while the lateral walls are made of Gd-loaded acrylic. All the TPC surfaces in contact with the active argon volume will be coated with a tetraphenyl-butadiene wavelength shifter (TPB) to convert LAr scintillation light (at 128 nm) to a wavelength detectable at high efficiency by the light detection planes, made of silicon photomultipliers (SiPMs). The lateral walls are coated, behind the reflector panels, with clevios electrodes, to maintain a uniform 200 V/cm drift field in the bulk. A transparent clevios coating is present also on the inner surfaces of top and bottom windows, to make anode and cathode of the drift cage. A steel mesh is positioned 1 cm below the top window, to generate the strong electric field required for the extraction of ionization electrons in the gaseous phase (7 mm thick).

2.3 The photo-detector systems

The light detection system is made of 2112 SiPM-based Photodetector Modules (PDMs), viewing the argon volume through the top and bottom windows of the acrylic vessel. The the light detection system must satisfy stringent performance requirements in order to meet the physics goals of the experiment. The choice of SiPMs is motivated by the numerous advantages of SiPMs compared to traditional photo-multipliers: the higher photo-detection efficiency (typically exceeding 50% at 420 nm), the superior single electron resolution, the lower operation bias voltage (< 100 V) and the smaller amount of required material, making them potentially extremely radio-pure. Several years of R&D within the Collaboration allowed to overcome the challenges related to this technology. These are: the coverage of large areas with small devices (individual SiPMs usually don't exceed the ~cm scale, due to the large capacitance per unit area responsible for slowing down the signal); and the high dark noise rate, solved by adopting the NUV-LF technology developed at FBK [11].

Each TPC PDM consists of 4 tiles of grouped SiPMs, covering a total area of 100×100 mm^2 operated as a single detector. Besides the tile, each module will also contain a cryogenic preamplifier board that will amplify and shape the signal in the immediate proximity of the sensor. In order to maximize the amplification factor while preserving a stable signal bandwidth and SNR, the Collaboration has developed and optimized a transimpedance amplifier with excellent performance at 87K. Figure 2 (left) shows one prototype assembly of 4 PDMs partially populated.

The proposed configuration matches all the performance requirements and the Collaboration is currently entering the mass production phase of these devices, foreseen at the NOA facility (Nuova Officina Assergi) at LNGS.

2.4 The radiogenic neutron veto

Neutrons are one of the most dangerous background to WIMP searches, as their signature can mimic that of the expected signal. Detailed simulations show that the combined use of Gd-loaded acrylic and an instrumented UAr volume surrounding the TPC, can reduce the radiogenic neutron background by a factor of ~ 10. Gadolinium is widely used for neutron tagging due to the high neutron capture cross section and the large amount of energy released after a neutron capture event (~ 9 MeV). The typical capture time of neutrons in the Gd-loaded acrylic is $\tau_c < 50\mu$s. The neutron background is efficiently suppressed if coincident energy deposits are observed in both the TPC and the surrounding UAr. The UAr volume surrounding the TPC is instrumented with 480 SiPM-based light detectors, distributed uniformly on the surfaces. The inner surfaces are covered with PEN sheets (acting as wavelength shifter) and Tyvek reflectors. This will enable to efficiently collect the scintillation lights from energy deposits in the UAr, with a light yield of about 2 PE/keV.

The Collaboration pioneered the development of Gd-loaded acrylic and identified, in combination with industrial partners, a strategy to produce 17 cm thick acrylic panels, with satisfactory Gd-loading uniformity (at 1% concentration) and meeting the required specifications in terms of radio-purity. These panels will be used to build the TPC barrel and the top and bottom cap of the neutron shield, to guarantee 4π shielding of the active volume.

3 Projected sensitivity

The sensitivity projections for DarkSide-20k, planned to start taking data in 2026, are show in Fig. 3. These are obtained with a Profile-Likelihood Ratio approach, where signal form spin-independent DM-nucleus interactions is compared to the expected backgrounds from radiogenic neutrons and coherent elastic neutrino-nucleus scattering (domainated by the contribution from atmospheric neutrinos). The electron recoil backgrounds (induced by β-decays and γ-ray interactions) are efficiently removed by the powerful PSD, which sets the acceptance to nuclear recoil backgrounds and signal. The projections are shown for two scenarios: one which only uses the events occurring in the innermost 20.2 t fiducial volume; and one which uses nearly the whole active volumes. The sum of NR backgrounds is expected to be

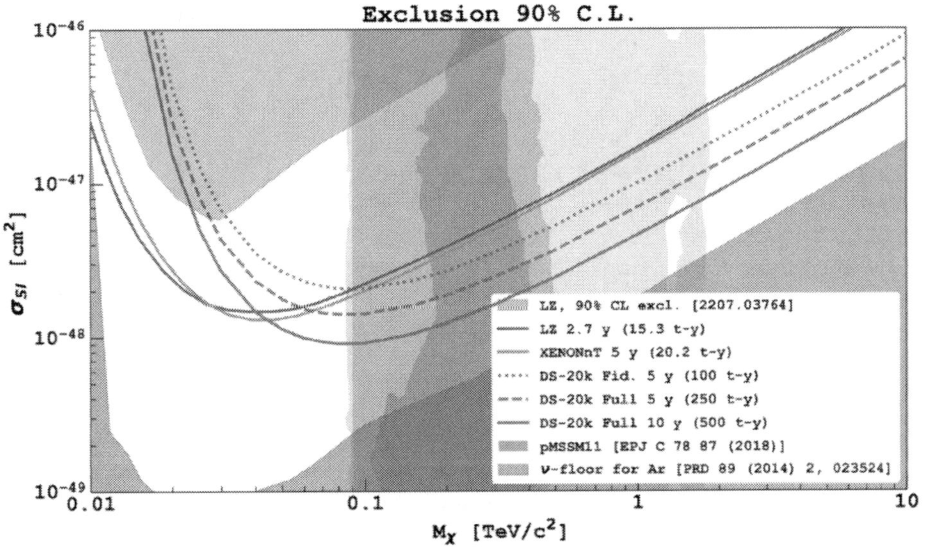

Figure 3. Projected 90% CL exclusion sensitivity for the Darkside-20k experiment for the spin-independent WIMP-nucleon cross section σ_{SI} as function of the WIMP mass M_χ, compared to existing constraints from LZ [12] and the projected sensitivities for XENON-nT [13] and LZ [14]. The shaded grey area represents the neutrino floor, while the teal shaded area show the parameter space favored by the pMSSM11 model [15].

< 0.1 events in a 5 yr exposure in the FV, while neutrinos will contribute to 1.7 ± 0.3 events. The full volume analysis accounts for higher backgrounds according to the weaker fiducialization, and it is shown for a 5 and a 10 yr lifetime run.

References

[1] R. Ajaj et al., Phys. Rev. D **100**, 022004, (2019)
[2] P. Agnes et al., Phys. Lett. B **743**, 456, (2015)
[3] P. Agnes et al., Phys. Rev. D **98**, 102006, (2018)
[4] P. Agnes et al., Phys. Rev. Lett. **121**, 081307, (2018)
[5] P. Agnes et al., Phys. Rev. Lett. **121**, 111303, (2018)
[6] P. Agnes et al., arXiv:2207.11966
[7] P. Agnes et al., Phys. Rev. D **93**, 081101, (2016)
[8] B. Abi et al., arXiv:1706.07081 (2017)
[9] P. Agnes et al., Eur. Phys. J. C **81**, 359, (2021)
[10] C.E. Aalseth et al, JINST **15**, P02024, (2020)
[11] F. Acerbi et al., IEEE Transactions on Electron Devices, **vol. 64**, 2, 521-526,(2017)
[12] D. S. Akerib et al., Phys. Rev. D **101**, 351 (2020)
[13] E. Aprile et al., JCAP 11 **031**, (2020)
[14] D. Akerib et al., Phys. Rev. D **101**, 351, (2020)
[15] E. Bagnaschi et al., Eur. Phys. J. C **78**, 8738 (2018)

EPJ Web of Conferences **280**, 06004 (2023) https://doi.org/10.1051/epjconf/202328006004
RICAP-22

Directionality for nuclear recoils in a LAr TPC

N. Pino[1,2,4,*], *P.* Agnes[5,20], *I.* Ahmad[6], *S.* Albergo[1,2,4], *I.*Albuquerque[7], *M.* Ave[7], *W. M.* Bonivento[8], *B.* Bottino[9,10], *M.* Cadeddu[8], *A.* Caminata[10], *N.* Canci[11], *G.* Cappello[1,2], *M.* Caravati[8,12], *S.* Catalanotti[14,15], *V.* Cataudella[14,15], *R.* Cesarano[16,17], *C.* Cicalò[8], *G.* Covone[14,15], *A.* de Candia[14,15], *G.* De Filippis[14,15], *G.* De Rosa[14,15], *D.* Dell'Aquila[3,32], *S.* Davini[10], *C.* Dionisi[16,17], *G.* Dolganov[18], *G.* Fiorillo[14,15], *D.* Franco[19], *C.* Galbiati[9,20], *M.* Gulino[3,21], *V.* Ippolito[17], *N.* Kemmerich[7], *M.* Kimura[6], *M.* Kuss[13], *M.* La Commara[22,14], *X.* Li[9], *S. M.* Mari[23,24], *C. J.* Martoff[25], *G.* Matteucci[15], *V.* Oleynikov[26], *M.* Pallavicini[27,10], *L.* Pandola[3], *M.* Rescigno[17], *J.* Rode[19,28], *S.* Sanfilippo[3,23,24], *A.* Sosa[7], *Y.* Suvorov[14,15], *G.* Testera[10], *A.* Tricomi[1,2,4], *M.* Wada[6], *H.* Wang[29], *Y.* Wang[30,31], and *P.* Zakhary[6]

[1]Dipartimento di Fisica e Astronomia, Università di Catania, Catania, Italy
[2]Istituto Nazionale di Fisica Nucleare Sezione di Catania, Catania, Italy
[3]Istituto Nazionale di Fisica Nucleare, Laboratori Nazionali del Sud, Catania, Italy
[4]Centro Siciliano di Fisica Nucleare e Struttura della Materia, Catania, Italy
[5]Department of Physics, University of Houston, Houston, USA
[6]AstroCeNT, Nicolaus Copernicus Astronomical Centre, Warsaw, Poland
[7]Instituto de Fìsica, Universidade de São Paulo, São Paulo, Brasil
[8]Istituto Nazionale di Fisica Nucleare Sezione di Cagliari, Cagliari, Italy
[9]Physics Department, Princeton University, Princeton, USA
[10]Istituto Nazionale di Fisica Nucleare Sezione di Genova, Genova 16146, Italy
[11]Istituto Nazionale di Fisica Nucleare, Laboratori Nazionali del Gran Sasso, Assergi, Italy
[12]Dipartimento di Fisica, Università degli Studi di Cagliari, Cagliari, Italy
[13]Istituto Nazionale di Fisica Nucleare Sezione di Pisa, Pisa, Italy
[14]Istituto Nazionale di Fisica Nucleare Sezione di Napoli, Napoli, Italy
[15]Dipartimento di Fisica, Università degli Studi Federico II, Napoli, Italy
[16]Dipartimento di Fisica, Sapienza Università di Roma, Roma, Italy
[17]Istituto Nazionale di Fisica Nucleare Sezione di Roma, Roma, Italy
[18]National Research Centre Kurchatov Institute, Moscow, Russian Federation
[19]APC, Université Paris Diderot, CNRS/IN2P3, Paris, France
[20]Gran Sasso Science Institute, L'Aquila, Italy
[21]Università di Enna Kore, Enna, Italy
[22]Dipartimento di Farmacia, Università degli Studi Federico II, Napoli, Italy
[23]Dipartimento di Matematica e Fisica, Università degli Studi Roma Tre, Roma, Italy
[24]Istituto Nazionale di Fisica Nucleare Sezione di Roma Tre, Roma, Italy
[25]Physics Department, Temple University, Philadelphia, USA
[26]Budker Institute of Nuclear Physics (BINP SB RAS), Novosibirsk, Russian Federation
[27]Dipartimento di Fisica, Università degli Studi di Genova, Genova, Italy
[28]LPNHE Paris, Université Pierre et Marie Curie, Université Paris Diderot, CNRS/IN2P3, Paris, France
[29]Physics and Astronomy Department, University of California, Los Angeles, USA
[30]Institute of High Energy Physics, Beijing, China
[31]University of Chinese Academy of Sciences, Beijing, China
[32]Università degli Studi di Sassari, Sassari, Italy

[*]e-mail: noemi.pino@ct.infn.it

Abstract. In the direct searches for Weakly Interacting Massive Particles (WIMPs) as Dark Matter candidates, the sensitivity of the detector to the incoming particle direction could provide a smoking gun signature for an interesting event. The SCENE collaboration firstly suggested the possible directional dependence of a dual-phase argon Time Projection Chamber through the columnar recombination effect. The Recoil Directionality project (ReD) within the Global Argon Dark Matter Collaboration aims to characterize the light and charge response of a liquid Argon dual-phase TPC to neutron-induced nuclear recoils to probe for the hint by SCENE. In this work, the directional sensitivity of the detector in the energy range of interest for WIMPs (20-100 keV) is investigated with a data-driven analysis involving a Machine Learning algorithm.

1 Introduction

Directionality should be a key aspect to flag a nuclear recoil (NR) event as a WIMP's one and a crucial tool to reject isotropic background sources, like the irreducible neutrino fog. In a dual-phase argon-based Time Projection Chamber (TPC) signals coming from two different processes are collected: prompt scintillation light (S1) in liquid and electroluminescence light (S2) in gas [1]. The presence of an electric field, the so-called drift field $\mathcal{E}_d$, allows electrons from ionization to avoid recombination. These electrons are drifted into the gas layer of the detector (gas pocket) and originate a delayed signal by electroluminescence. S2-S1 correlation could in principle be used in order to infer information about the initial direction of the nuclear recoils (NRs) with respect to $\mathcal{E}_d$ exploiting the columnar recombination effect [2, 3].

2 The ReD Experiment

The first hint of a direction-sensitive effect in the response of a liquid argon (LAr) TPC comes from the SCENE experiment [4]. To scrutinize it, the DarkSide collaboration, within the Global Argon Dark Matter Collaboration (GADMC), undertook a dedicated project, the Recoil Directionality (ReD) experiment [1]. Therefore, a compact, cubic-shape LAr TPC (volume of 150 cm^3) has been built and characterized to evaluate the performance of the detector system in the same energy range as SCENE[1]. Furthermore, ReD is the first experiment with a LAr TPC equipped with new-generation cryogenic Silicon Photomultipliers (SiPMs). The TPC was irradiated for 14 days with quasi-mono energetic neutrons at INFN LNS, in Catania. Neutrons directed toward the detector are produced via the p(^{7}Li,^{7}Be)n reaction, and selected by tagging the accompanying ^{7}Be ions in a $\Delta E/E$ Silicon telescope. A neutron spectrometer, made by 7 Liquid Scintillators (LSci), detects downstream neutrons from NR at $E_R = 72$ keV, thus closing the (n, n') kinematics and allowing to derive also the direction of the recoiling Ar nucleus. The experimental setup (Fig. 1 (a)) is conceived to select events of NR in the TPC at a chosen recoil energy E_R within the range of interest for WIMPs, but with momenta at different angle θ_r with respect to $\mathcal{E}_d$. The scrutinized recoil directions θ_r by the neutron spectrometer are 0°, 20°, 40° and 90°. The golden-plated events, for which θ_r is reported, are those passing the selection of the tagger detectors (Si-telescope, TPC and one LSci) and a series of quality cuts. Those events are referred to as triple coincidence events.

3 Models for Recombination

The electron-ion pairs formed in the liquid target after ionization can recombine, thus contributing to the S1 signal at the expense of S2. However, under the presence of $\mathcal{E}_d$, the ionization electrons can be extracted from the interaction site and drifted towards the volume of the TPC up to the gas pocket. Along their travel, if the ionization track is parallel to the field [1], electrons will pass through the ionization column and will have a higher probability to recombine with Ar ions. On the contrary, a higher S2 signal is expected for momenta perpendicular to the field. Therefore, a proper modeling of the electron-ion cloud is needed to investigate the sensitivity of the directional response of a LAr-TPC through the S2-S1 correlation. One of the commonly-used models is the Thomas-Imel one [5]. It does not capture the directional sensitivity since it describes the cloud as a uniform cubic volume. On the other hand, the columnar recombination model introduced by Jaffé [2] overestimates the directional response since it treats the ionization cloud as an infinitely long cylinder. The ReD data are finally analyzed following a novel model that describes the cloud as an elongated ellipsoid with a single adimensional parameter R related to the non-sphericity of the initial electron-ion cloud [6, 7]. Starting from the equations describing the evolution of the electron-ion cloud, the dependence of the recombination fraction upon the angle between the track and drift field is described as $f(R, \theta_r) = \sqrt{\sin^2\theta_r + \cos^2\theta_r/R^2}$. An R value higher than 1 stands for a net directional effect, while for $R = 1$ any directional dependence is canceled and the Thomas-Imel configuration is restored.

4 A Data-Driven Analysis with Machine Learning

The absence of a well-established model allows for alternative approaches that do not require a theory as a starting point to describe a phenomenon. Artificial Intelligence techniques and Machine Learning (ML) are well suited to extrapolate the trend of a phenomenon starting from patterns and correlations in a dataset. In parallel with the statistical analysis carried out with an unbinned maximum likelihood fit [8] according to the model of [6], a data-driven analysis is then performed using ML. The strategy aims to train a model using numerical features extracted from NR events produced in the ReD TPC, thus starting from the assumption that the angle between the momentum of the recoiling Ar nucleus and the electric field is not relevant for the recombination. If it were true, the effect of recombination would not have privileged regions of θ_r and it could be uniformly distributed among the events of the dataset. Otherwise, if a deviation from the hypothesized behavior were found, one could indirectly prove the recombination dependence of the recoil angle. The derived model aims to extract a pattern from a sample of NR events including all θ_r angles, thus becoming capable of predicting the value of the ionization signal. Therefore, the S2 signal for each event will be predicted by using only the spatial coordinates of the event and the scintillation signal. The vector of features passed as an input is composed of the scintillation signal S1 and the (x,y,z) position of the ionization signal. The model should be able to predict the value of the S2 corrected for TPC non-uniformity effects. The problem is a supervised non-linear regression one since the model is trained on the vector of features as an input and the measured S2 signal as the target output. The chosen algorithm is the Extreme Gradient Boosting one (`xgboost`), well known in the literature for its good performance and multi-field-possible application to various problems [9]. To evaluate the accuracy of the derived model, the relative prediction error for each event is calculated, ϵ_{pred}, defined as

$$\epsilon^i_{pred} = \frac{S2^i_{measured} - S2^i_{predicted}}{S2^i_{measured}} \quad (1)$$

At the end of the training phase, the derived model was able to predict the value of the S2 signal for the events in the testing set with relative errors ϵ_{pred} approximately Gaussian-distributed with mean 0.0043(6) and standard deviation 0.09. Once the model is derived and tested it is used to make predictions on the triple coincidence dataset. First, ϵ_{pred} is calculated event-by-event and the final data sample from all events is divided into four subsets, one for each investigated θ_r, thus obtaining four ϵ_{pred} distributions. The mean value and the error on the mean are calculated for each distribution (Fig. 1 (b)). It is possible to note that the ϵ_{pred}

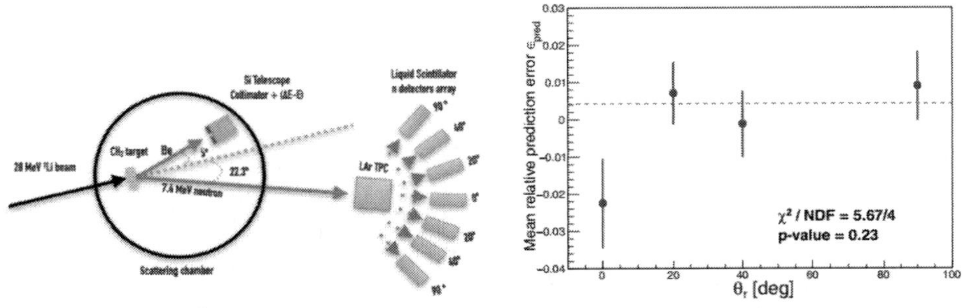

(a) *Setup of ReD* (b) ϵ_{pred} *distributions from triple-coincidences*

Figure 1: (a) Schema of the experimental setup of the ReD experiment. (b) Mean relative prediction error for each ϵ_{pred} distribution obtained by splitting data into the four angular datasets. The red dashed line marks the $\epsilon_{pred} = 0.0043$ level (no directional effect).

value is on average lower for those events with traces parallel to the direction of the drift field. From the definition in (1), this should stand for a predicted S2 value greater than the experimentally measured one. This shows that the derived model tends to overestimate S2 for events with traces parallel to $\mathcal{E}_d$. Such a result is expected in the case of directionality effects, as traces parallel to $\mathcal{E}_d$ would result in enhanced S1 signals and reduced S2. The statistical significance of this result is evaluated by performing a χ^2 test: the p-value calculated for the null hypothesis of no directional effect is about 23%.

5 Conclusion

The directional sensitivity of the ReD TPC is investigated with two complementary approaches, one according to a direction-dependent LAr charge recombination model with a likelihood statistical analysis [7, 8], and another one, described here, developing a data-driven analysis with ML techniques. Even if the observed difference in the behavior of the detector for NRs parallel to the $\mathcal{E}_d$ could possibly hint at a directional sensitivity, the result of the test statistic does not support it within a significant confidence level.

References

[1] P. Agnes et al., Eur. Phys. J. C **81**, 1014 (2021)
[2] G. Jaffè, Annalen Phys. **347**, 303 (1913)
[3] J.B. Birks, Proc. Phys. Soc. **A64**, 874 (1951).
[4] H. Cao et al. (The SCENE Collaboration), Phys. Rev. D **91**, 092007 (2015)
[5] J. Thomas, D.A. Imel, Phys. Rev. A **36(2)**, 614 (1987)
[6] V. Cataudella et al., JINST **12**, P12002 (2017)
[7] X. Li, talk at Int. Conf. on Topics in Astroparticle and Underground Physics, 2021
[8] S. Sanfilippo et al., IOP-Journal of Physics: Conference Series (in press), (2022)
[9] T. Chen et al., R package version 0.4-2 **1.4**, 1-4 (2015)

Pierre Auger Observatory and Super Heavy Dark Matter

Roberto Aloisio[1,2,*] for the Pierre Auger Collaboration[3,**]

[1]Gran Sasso Science Institute, viale F. Crispi 7, 67100 L'Aquila, Italy
[2]INFN - Laboratori Nazionali Gran Sasso, via G. Acitelli 22, 67100 Assergi (AQ), Italy
[3]Observatorio Pierre Auger, Av. San Martín Norte 304, 5613 Malargüe, Argentina

Abstract. We briefly discuss the connection of the Pierre Auger Observatory data with a large class of dark matter models based on the early universe generation of super heavy particles, their role in the solution of the dark matter problem, highlighting the remarkable constraining capabilities of the Auger observations.

1 Introduction

Ultra-high energy cosmic rays (UHECR) are the most energetic particles ever observed with energies larger than 10^{20} eV. These extreme energies, as high as 10^{11} GeV, eleven orders of magnitude above the proton mass and "only" eight below the Planck mass, are a unique workbench to probe new ideas, models and theories beyond the Standard Model (SM) of particle physics, which show their effects at energies much larger than those ever obtained, or obtainable in the future, in accelerator experiments. This is the case of theories with Lorentz invariance violations [1–4] or models of Dark Matter (DM) with super heavy particles [5? –10], that connect UHECR observations with the Dark Sector (DS) and the cosmological evolution of the early universe.

The leading paradigm to explain DM observations is based on the Weakly Interactive Massive Particle (WIMP) hypotheses, a stable DM particle with mass in the range of $0.1 \div 10$ TeV [11]. Searches for WIMPs provided no evidence of a clear candidate and alternative solutions to the DM problem should be considered. An alternative to WIMP models is represented by the scenarios based on long lived super-heavy relics, that can be produced by several mechanisms taking place during the inflationary phase or just after, in the re-heating phase [5–7, 9, 10]. Once created in the early universe, the abundance of the long-lived super-heavy particles can evolve to match the DM density observed today, the so-called Super Heavy Dark Matter (SHDM). This conclusion can be drawn under three general hypotheses: (i) SHDM in the early universe never reaches local thermal equilibrium; (ii) SHDM particles have mass M_X of the order of the inflaton mass or higher; and (iii) SHDM particles are long-living particles with a lifetime exceeding the age of the universe, $\tau_X \gg t_0$.

By far the largest experiment devoted to the observation of UHECR is the Pierre Auger Observatory (Auger) in Argentina. UHECR can be observed only indirectly through the detection of their interaction products with the Earth's atmosphere. An UHE particle interacting

*roberto.aloisio@gssi.it
**spokespersons@auger.org.
A complete list of authors can be found at https://www.auger.org/archive/authors_2022_09.html

with a nucleus of the atmosphere produces both hadronic and electromagnetic cascades of particles, collectively called Extensive Air Shower (EAS). In Auger the EAS observation is performed by detecting the EAS particles that reach the ground and by the (coincident) detection of the fluorescence emission produced in the atmosphere, the so-called hybrid events.

As always in the case of cosmic rays, the information we can gather experimentally concerns: energy spectrum, mass composition and arrival direction anisotropy. The Auger observations of UHECR clarified several important facts [12–15, 17]: (i) UHECR are charged particles, with severe upper limits on UHE photon and neutrino fluxes (see later); (ii) the spectrum observed at the Earth shows a slight flattening at energies around 5×10^{18} eV (called the ankle) with (iii) an instep at 10^{19} eV and (iv) a steep suppression at the highest energies; (v) mass composition is dominated by light particles (proton and helium) at energies around 10^{18} eV becoming progressively heavier starting from energies around 3×10^{18} eV, with lacking of light particles toward the highest energies; (vi) at energies larger than 8×10^{18} eV UHECR show a strong signal of dipole anisotropy in the arrival directions, with 6% amplitude and the phase pointing toward the direction of the galactic anti-center.

In section 2, we will briefly introduce SHDM models discussing their connection to the Auger observations and the related constraining capabilities. We conclude in section 3.

2 Super-heavy dark matter models and constrains

Recently, using the LHC measurements of the masses of the Higgs boson and Top quark, it has been pointed out that the Higgs potential is stable until very high energy scales $\Lambda_I \simeq 10^{10} \div 10^{12}$ GeV and it might be possible to extrapolate the SM to even higher energies up to the Planck mass $M_P = 10^{19}$ GeV [16]. This fact implies that the mass spectrum of the DS is restricted to ultra-high energies and SHDM could play a role. In the following, we will consider SHDM with masses in the range between 10^8 GeV up to the Planck mass.

Apart from the gravitational interaction, SHDM can be coupled to ordinary matter also through some super-weak coupling, driven by a high energy scale $\Lambda > \Lambda_I$ as the Grand Unification Scale ($\Lambda_{GUT} \simeq 10^{16}$ GeV). To assure long-lived particles, the interaction SHDM-SM should be suppressed by some power n of the high energy scale Λ, with a decay SHMD $\rightarrow$ SM characterised by a lifetime $\tau_X \simeq (M_X \alpha_{X\Theta})^{-1}(\Lambda/M_X)^{2n-8}$, being $\alpha_{X\Theta}$ the reduced coupling constant between SHDM and SM particles [5]. In the case in which SHDM interacts with SM particles only through the gravitational interaction[1], assuming the very general case of a DS characterised by its own non-abelian gauge symmetry, SHDM can decay only through non-perturbative effects as the instanton-induced decay [5]. In this case, the lifetime of SHDM follows from the corresponding instanton transition amplitude that, being exponentially suppressed, provides long-living particles. Considering the zeroth order contribution, the instanton-induced lifetime can be written as $\tau_X \simeq M_X^{-1} \exp(4\pi/\alpha_X)$, being α_X the reduced coupling constant of the hidden gauge interaction in the DS.

Under very general assumptions [5? –10], we can determine the composition and spectra of the standard model particles produced by the SHDM decay. Typical decay products are couples of quark and anti-quark[2] that, through a cascading process (jets), give rise to neutrinos, gamma rays and nucleons. As follows using the DGLAP equations [7], these particles exhibit a flat spectrum, that, as shown in the left panel of figure 1, at the relevant energies, can be approximated as $dN/dE \propto E^{-1.9}$, independently of the particle type, with a photon/nucleon ratio of about $\gamma/N \simeq 2 \div 3$ and a neutrino nucleon ratio $\nu/N \simeq 3 \div 4$, quite independent of the energy range [7]. Therefore, the most constraining limits on the SHDM mass and lifetime

[1]Note that, in this case, a SHDM particle can be also called Planckian-Interacting Massive Particle (PIDM).
[2]For a discussion of alternative (leptonic) decay patterns see [8].

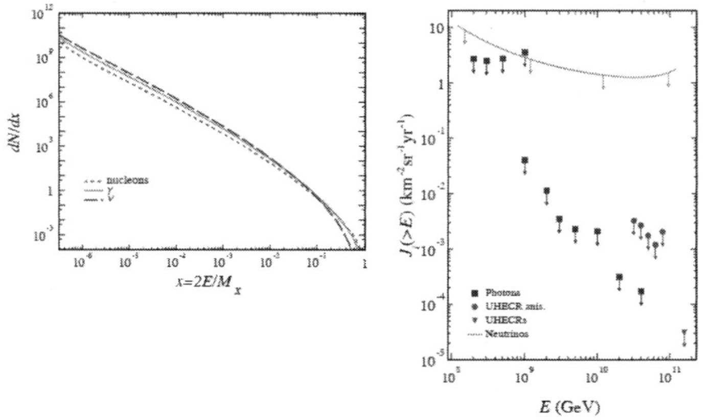

Figure 1. [Left Panel] Distribution function of the SM products in the SHDM decay. [Right Panel] Auger experimental limits on the fluxes of gamma rays and neutrinos. Figures taken from [5].

are those coming from the (non) observation of UHE photons and, even to a lesser extent, neutrinos.

Photons with energies larger than 10^8 GeV have an absorption length roughly of the size of our galaxy, increasing up to the scale of the local group ($1 \div 10$ Mpc) at energies larger than 10^{10} GeV. Moreover, as follows from numerical simulations of structure formation, DM is clustered in galactic halos with an over-density at the level of 10^5 [7], it implies that only UHE photons from the decay of SHDM in our our own galaxy play a role. The observation of photons with energy larger than 10^8 GeV is affected by the background of cosmogenic photons, i.e. UHE diffuse photons produced by the propagation of UHECR. This background depends on the mass composition of UHECR and on the details of their astrophysical sources, it can be estimated to range (at most) around 10^{-2} km^{-2} sr^{-1} yr^{-1} above 10^8 GeV and around $10^{-3.5}$ km^{-2} sr^{-1} yr^{-1} above 10^9 GeV [17]. These cosmogenic diffuse fluxes are two orders of magnitude below the sensitivity of Auger and can be safely neglected. The search for UHE photon events in the Auger dataset exploits the well known differences in the EAS initiated by photons or hadrons: EAS generated by photons develop deeper in the atmosphere and are characterised by a reduced number of muons, with a substantial detectable difference in the lateral distribution of EAS particles [17]. In the right panel of figure 1 we plot the very stringent limits on the photon and neutrino fluxes as extracted from the Auger data (2022 update) in the energy range $10^8 \div 10^{11}$ GeV [5, 6, 17].

Assuming a DM density in our galaxy as in the Navarro-Frenk-White profile [18], we can determine the expected flux of UHE photons produced by the decay of SHDM. Requiring that this flux does not exceed the Auger experimental limits, it is possible to constrain the mass and lifetime of SHDM. In the left and central panels of figure 2, we plot the limits on M_X and the coupling SHDM-SM $\alpha_{X\Theta}$, in the case of direct coupling (left panel), and the coupling of the hidden DS gauge interaction α_X, in the case of the instanton-induced decay (central panel). Assuming the class of models based on SHDM creation during the reheating phase, the limits on M_X and τ_X can be rewritten in terms of the cosmological parameters characterising the generation mechanism of SHDM. In the right panel of figure 2, we plot

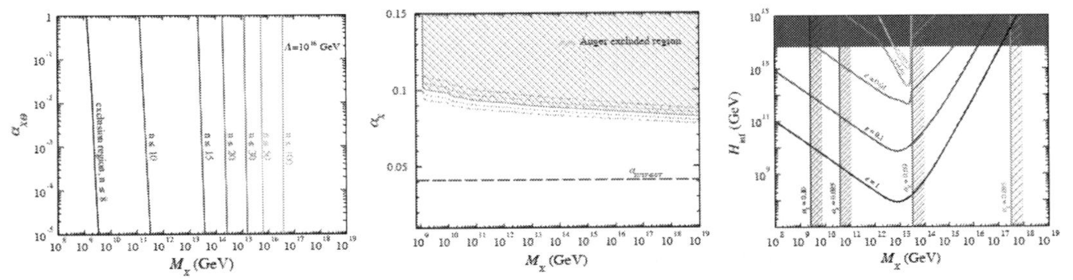

Figure 2. Limits fixed by the Auger observations on the SHDM mass M_X and: (left panel) the coupling SHDM-SM $\alpha_{X\Theta}$ for different values of the suppression scale order n, (central panel) the coupling of the hidden DS gauge interaction α_X in the case of the instanton-induced decay, (right panel) the universe expansion rate at the end of inflation H_{inf}. In the right panel, we plot the limits for different values of the reheating efficiency ϵ (see text). Figures taken from [5].

the limits coming from Auger data on the age of the universe at the end of inflation H_{inf}^{-1} for different values of the reheating efficiency $\epsilon = (\Gamma_\phi/H_{inf})^{1/2}$, being $\Gamma_\phi = g_\phi^2 M_\phi/8\pi$ the inflaton decaying amplitude into SHDM (M_ϕ inflaton mass; g_ϕ inflaton coupling with SHDM) [5, 6].

3 Conclusions

We conclude by stating the importance of Auger observations on UHE photons and neutrinos, as they enable to constrain models of new physics connected to the grand unification scale and the reheating phase of the universe. This is the case of SHDM that, as discussed above, being a viable alternative to the WIMP paradigm for DM, can be tested only through UHECR observations, eventually connected with cosmological observations.

References

[1] P. Abreu *et al.* [Pierre Auger], JCAP **01**, no.01, 023 (2022) [arXiv:2112.06773].
[2] A. Addazi, *et al.* Prog. Part. Nucl. Phys. **125**, 103948 (2022) [arXiv:2111.05659].
[3] R. Aloisio, P. Blasi, *et al.* Phys. Rev. D **62**, 053010 (2000) [arXiv:astro-ph/0001258].
[4] R. Aloisio, *et al.* Astropart. Phys. **19**, 127-133 (2003) [arXiv:astro-ph/0205271].
[5] P. Abreu *et al.* [Pierre Auger], [arXiv:2208.02353].
[6] P. Abreu, *et al.* [Pierre Auger], [arXiv:2203.08854].
[7] R. Aloisio, *et al.* Phys. Rev. D **74**, 023516 (2006) [arXiv:astro-ph/0604311].
[8] C. Guépin, *et al.* Phys. Rev. D **104**, no.8, 083002 (2021) [arXiv:2106.04446].
[9] R. Aloisio, *et al.* JCAP **08**, 024 (2015) [arXiv:1504.01319].
[10] R. Aloisio, *et al.* Astropart. Phys. **29**, 307-316 (2008) [arXiv:0706.3196].
[11] L. Bergström, Rept. Prog. Phys. **63**, 793 (2000) [arXiv:hep-ph/0002126].
[12] A. Coleman, *et al.* [arXiv:2205.05845].
[13] R. Alves Batista, *et al.* Front. Astron. Space Sci. **6**, 23 (2019) [arXiv:1903.06714].
[14] R. Aloisio, *et al.* Springer Cham 2018, [arXiv:1707.06147].
[15] R. Aloisio, PTEP **2017**, no.12, 12A102 (2017) [arXiv:1707.08471].
[16] D. Buttazzo, *et al.* JHEP **12**, 089 (2013) [arXiv:1307.3536].
[17] P. Abreu *et al.* [Pierre Auger], Astrophys. J. **933** (2022) no.2, 125 [arXiv:2205.14864].
[18] J. F. Navarro, *et al.* Astrophys. J. **462** (1996), 563-575 [arXiv:astro-ph/9508025].

The GAPS experiment: low-energy antinuclei measurements for dark matter searches

N. Marcelli[1,*] on behalf of the GAPS collaboration[2]

[1]INFN, Sezione di Trieste, Padriciano 99, I-34149 Trieste, Italy
[2]Full author list: https://gaps1.astro.ucla.edu/gaps/authors/

Abstract. GAPS (General Anti-Particle Spectrometer) is a balloon-borne experiment designed to measure low-energy (<0.25 GeV/n) cosmic antinuclei (i.e., antiprotons, antideuterons, and antihelium nuclei) as a signature of dark matter annihilation or decay. According to viable beyond-the-Standard Model theories, the predicted dark matter signal in the low-energy antideuterons and antihelium nuclei channels is several orders of magnitude higher than the astrophysical background. The experiment will conduct a series of at least three long-duration balloon flights at high altitudes from Antarctica. The instrument is composed of a Si(Li) tracker surrounded by a Time-of-Flight system made of plastic scintillators. GAPS uses the novel exotic-atom detection technique in which an antinucleus is captured by the tracker material and forms an exotic atom. This excited exotic atom decays within the order of nanoseconds emitting X-rays at specific energies defined by the atomic transitions and annihilates emitting secondary particles (mainly pions and protons). The measured quantities (e.g., dE/dx, time of flight, annihilation vertex position, X-rays energies, etc.) allow for identifying antinuclei with high precision.

1 Introduction

The General AntiParticle Spectrometer (GAPS) [1, 2] is the first experiment devised to measure low-energy antinuclei, exploring the energy region below 0.25 GeV/n. According to viable beyond-the-Standard Model theories, the predicted dark matter (DM) signal in the low-energy antideuterons and antihelium nuclei channels is several orders of magnitude higher than the astrophysical background [3, 4]. In this scenario, GAPS will operate in an almost background-free energy region. The experiment will use a novel detection technique based on the formation and decay of exotic atoms. GAPS is planned to have at least three long-duration balloon (LDB) flights from McMurdo station in Antarctica. The first flight is expected to be performed in the austral summer of 2023-2024. GAPS will provide unprecedented sensitivity for antideuteron and antihelium nuclei [5, 6]. Moreover, GAPS will perform high statistic measurements of antiprotons and extend the energy coverage in an unexplored low-energy region [7].

*e-mail: nadir.marcelli@ts.infn.it

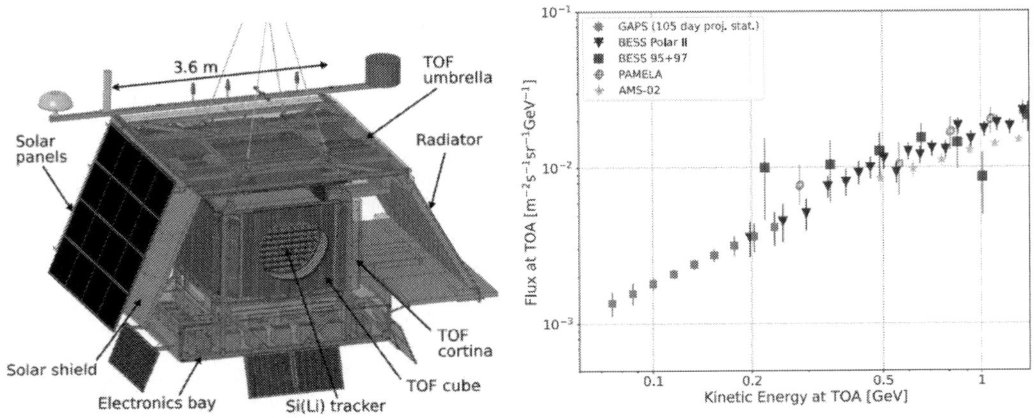

Figure 1. Left panel: A schematic view of the GAPS detector. **Right panel**: The GAPS sensitivity for cosmic antiproton spectrum (red) at the top of the atmosphere is shown with the statistics expected from three 35-day flights [7]. Data from BESS (1995 and '97 [12]), BESS Polar II [13], PAMELA (2006–09 [14]), and AMS-02 (2011–18 [15, 16]) are also shown.

2 The GAPS detector

Fig. 1 (left panel) shows a schematic view of the GAPS apparatus consisting of a Time-of-Flight (ToF) and tracking systems.

The ToF system consists of ~160 plastic scintillator paddles [8] and is arranged in an outer Tof system and an inner Tof system. The outer ToF consists of an umbrella of scintillator oriented horizontally and a cortina of four walls of scintillator oriented vertically. The inner ToF is a cube of scintillator paddles, consisting of four sides, a top and a bottom. The ToF system measures the time information necessary to reconstruct the velocity of particles and the ionization energy losses dE/dx of particles. The ToF also provides the overall trigger for GAPS.

The core of the apparatus is a tracking system made of ten planes of 12×12 cylindrical Si(Li) detectors each [6, 9]. For the first flight, ~1000 Si(Li) detectors will be equipped. On each supporting aluminum plane, the Si(Li) cylinders are arranged in a 6×6 array of modules, each with four Si(Li) detectors read out by a dedicated ASIC [10]. The Si(Li) detectors have a diameter of ~10 cm, are ~2.5 mm thick and are segmented into eight strips of equal area. Finally, the tracking system is kept at its operational temperature of ~-40° C by an oscillating heat pipe (OHP) thermal system [11].

3 Detection technique

GAPS uses a novel technique for antinuclei detection, based on the formation and decay of exotic atoms. The incoming primary antinucleus slows down due to ionization energy losses in the materials of the apparatus and, after losing all kinetic energy, can form an exotic atom by replacing a shell electron in a silicon or aluminum atom. During the de-excitation process, the exotic atom emits characteristic X-rays with energies determined by atomic transitions [5]. In the end, the primary particle annihilates with the target nucleus producing secondary particles, mainly pions and protons from a common vertex. The measurements of the time-of-flight, ionization losses dE/dx, and X-ray energies together with the reconstructed topology of the "annihilation star", allow to clearly identify antinuclei.

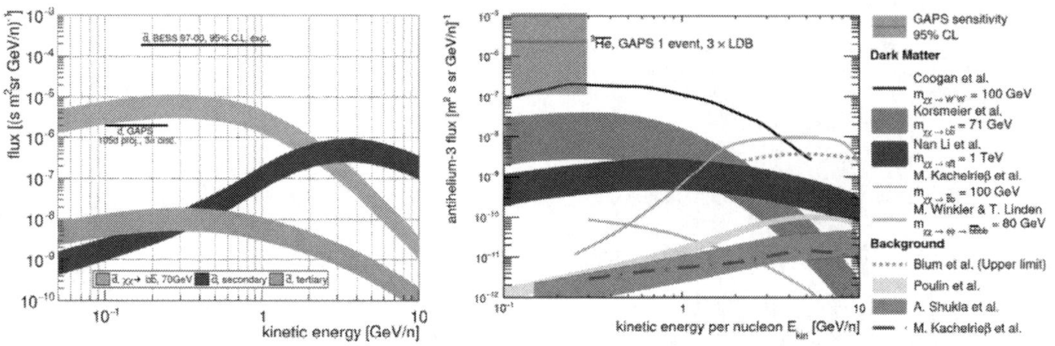

Figure 2. Left panel: the predicted antideuteron flux from a generic 70-GeV WIMP annihilating via $b\bar{b}$-pairs (cyan), as well as the predicted secondary (blue) and tertiary (pink) astrophysical antideuteron flux. The anticipated sensitivity of GAPS [5, 19] for a 3 σ discovery and the BESS 97–00 95% C.L. exclusion limits are shown [20]. **Right panel**: GAPS antihelium sensitivity (red line, with 95% C.L.) compared with the flux predicted by different dark matter and astrophysical background models [6].

In order to reconstruct the antinucleus annihilation topology, a custom algorithm was developed. First, the primary track is identified, followed by the secondary tracks associated with the primary track. Finally, the annihilation vertex is defined as the point which minimizes the distance between primary and secondary reconstructed tracks [17].

4 Antinuclei sensitivity

The sensitivity curves for antinuclei were determined using the GAPS simulation software, which reproduces the instrument geometry and materials using the Geant4 toolkit [18]. Moreover, a digitization process was applied to the simulated data to mimic a realistic instrument response for time, energy, and position measurements.

Fig. 1 (right panel) shows the expected antiproton measurement at the top of the atmosphere (TOA) for three ~35 day antarctic flights at ~37 km float altitude [7]. The GAPS precision measurement of antiprotons in this unexplored low-energy range will have about two orders of magnitude more low-energy antiprotons than BESS [12, 13] and PAMELA [14] combined, i.e., ~500 antiprotons expected for each flight. These measurements will provide strong constraints on DM candidates with masses <30 GeV and primordial black hole evaporation on Galactic length scales [19].

Antideuteron nuclei have never been observed in cosmic rays, and any detection would be sufficient to claim the discovery of new physics. Fig. 2 left pannel shows the GAPS sensitivity to low-energy cosmic antideuteron after three LDB flights, compared with the antideuteron fluxes expected from a generic 70-GeV WIMP annihilating into antideuterons via $b\bar{b}$-pairs (cyan) [5, 19]. As described in Sec. 1, the astrophysical background (blue and pink) is at least two orders of magnitude below the expected DM signal [21].

The flux of antihelium nuclei predicted by several dark matter models is illustrated in Figure 2, right panel. Three GAPS LDB flights will provide enough sensitivity to low-energy antihelium-3 nuclei, with the potential to probe dark matter models annihilating into W^+W^- [6].

Due to its novel detection technique and sensitivity to the lower-energy range, where the predicted contribution from most of the new-physics models is highest, GAPS will provide complementary measurements to the experiments operating so far.

Acknowledgements

This work is supported in the U.S. by the NASA APRA program (Grant Nos. NNX17AB44G, NNX17AB46G, and NNX17AB47G), in Japan by the JAXA/ISAS Small Science Program FY2017, and in Italy by Istituto Nazionale di Fisica Nucleare (INFN) and the Italian Space Agency (ASI) through the ASI INFN agreement No. 2018-22-HH.0: "Partecipazione italiana al GAPS - General AntiParticle Spectrometer." H. Fuke is supported by JSPS KAKENHI grants (JP17H01136, JP19H05198, and JP22H00147) and Mitsubishi Foundation Research Grant 2019-10038. The contributions of C. Gerrity are supported by NASA under award No. 80NSSC19K1425 of the Future Investigators in NASA Earth and Space Science and Technology (FINESST) program. R. A. Ong receives support from the UCLA Division of Physical Sciences. K. Perez and M. Xiao are supported by Heising-Simons award 2018-0766. F. Rogers is supported through the National Science Foundation Graduate Research Fellowship Program under Grant No. 1122374. Y. Shimizu receives support from JSPS KAKENHI grant JP20K04002 and Sumitomo Foundation Grant No. 180322. M. Yamatani receives support from JSPS KAKENHI grant JP22K14065. K. Yee is supported through the National Science Foundation Graduate Research Fellowship under grant 2141064.

References

[1] K. Mori et al., Astrophys. J. **566**, 604 (2002), astro-ph/0109463
[2] C.J. Hailey, New J. Phys. **11**, 105022 (2009)
[3] F. Donato et al., Phys. Rev. D **62**, 043003 (2000)
[4] N. Fornengo et al., Journal of Cosmology and Astroparticle Physics **2013**, 031 (2013)
[5] T. Aramaki et al., Astroparticle Physics **74**, 6 (2016)
[6] N. Saffold et al., Nucl. Instrum. Meth. A **997**, 165015 (2021), 2102.06168
[7] F. Rogers et al., Astroparticle Physics **145**, 102791 (2023)
[8] S. Quinn, PoS **ICRC2019**, 128 (2019)
[9] M. Kozai et al., Nucl. Instrum. Meth. A **947**, 2695 (2019), 1906.05577
[10] V. Scotti et al., PoS **ICRC2019**, 136 (2019)
[11] H. Fuke et al., in *2019 IEEE Nuclear Science Symposium and Medical Imaging Conference (NSS/MIC)* (2019), pp. 1–3
[12] S. Orito et al., Physical Review Letters **84**, 1078 (2000), astro-ph/9906426
[13] K. Abe et al., Physics Review Letters **108**, 051102 (2012), 1107.6000
[14] O. Adriani et al., JETP Letters **96**, 621 – 627 (2013)
[15] M. Aguilar et al. (AMS Collaboration), Phys. Rev. Lett. **117**, 091103 (2016)
[16] M. Aguilar et al., Physics Reports **894**, 1 (2021), the Alpha Magnetic Spectrometer (AMS) on the International Space Station: Part II - Results from the First Seven Years
[17] R. Munini et al., Astroparticle Physics **133**, 102640 (2021)
[18] S. Agostinelli et al., Nuclear Instruments and Methods in Physics Research Section A: Accelerators, Spectrometers, Detectors and Associated Equipment **506**, 250 (2003)
[19] P. von Doetinchem et al., Journal of Cosmology and Astroparticle Physics **2020**, 035 (2020)
[20] H. Fuke et al., Phys. Rev. Lett. **95**, 081101 (2005)
[21] M. Korsmeier, F. Donato, N. Fornengo, Phys. Rev. D **97**, 103011 (2018), 1711.08465

Primordial Black Holes as dark matter candidates: Constraints from the diffuse MeV emission

Joanna Berteaud[1,*], *Francesca* Calore[1], *Joaquim* Iguaz[1], *Pasquale Dario* Serpico[1], and *Thomas* Siegert[2]

[1]LAPTh, CNRS, USMB, F-74940 Annecy, France

[2]Institut für Theoretische Physik und Astrophysik, Universität Würzburg, Campus Hubland Nord, Emil-Fischer-Str. 31, 97074 Würzburg, Germany

Abstract. For more than 20 years, the Compton telescope had provided the best measurements of the Galactic diffuse MeV spectrum. Recently, our analysis of 16 years of data from the SPectrometer on INTEGRAL (SPI) measured this emission with a higher signal-to-noise ratio. At MeV energies, the dominant contribution to the diffuse emission comes from inverse Compton scattering. Nonetheless, sub-dominant emission from Primordial Black Hole (PBH) Dark Matter (DM) can be searched for in these data. Hypothetically formed from the collapse of over-densities before Big Bang nucleosynthesis, PBHs are interesting candidates for DM in the ΛCDM model of cosmology. PBHs of masses between 10^{16} and 10^{18} g, in the so-called asteroid mass range, are currently unconstrained and can saturate the DM cosmological abundance. MeV emission from PBH in this mass range is expected to come from PBH evaporation. We searched for the PBH signal with 16 years of SPI data, and demonstrated that PBHs cannot account for all the DM if their mass is smaller than 4×10^{17} g.

1 Introduction

Black holes (BHs) are objects so compact that nothing can escape from their gravitational influence. Unlike stellar BHs that form from the gravitational collapse of stars in the late Universe, primordial BHs (PBHs) are thought to be formed from over dense regions in the early universe, before Big Bang nucleosynthesis. Therefore, they could contribute to the dark matter (DM), which accounts for about 27% of the energy density of the Universe according to the ΛCDM model of cosmology

Figure 1 shows the existing constraints on f_{PBH}, the fraction of DM in the form of PBHs, as a function of M_{PBH}, the monochromatic PBH mass. For masses larger than 10^{22} g or smaller than 10^{17} g, PBH DM is well constrained by various observations but between these two values, a mass window remains where PBHs could still saturate the DM cosmological abundance. The lower part of this mass window, the asteroid mass range, can be probed by looking for the Hawking evaporation signal of PBHs in Galactic diffuse MeV data.

*e-mail: μberteaud@lapth.cnrs.fr

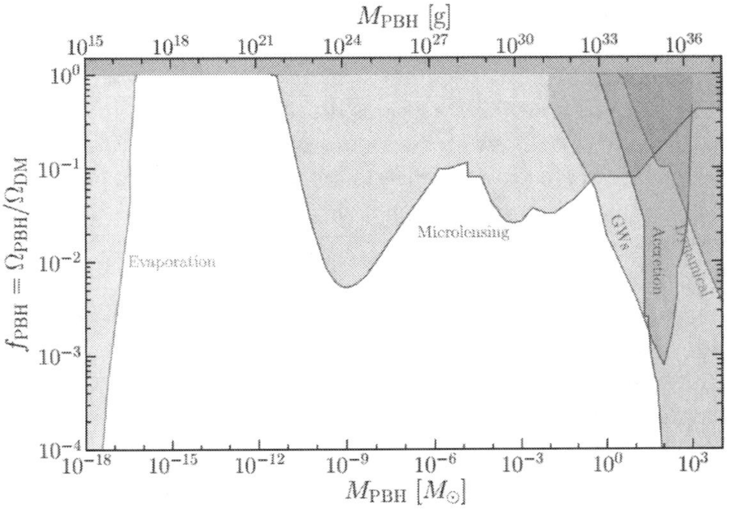

Figure 1. Constraints on PBH DM [1]. The possible fraction of DM in the form of PBHs $f_{\rm PBH}$ depends on the PBH mass $M_{\rm PBH}$. The shaded regions are excluded by various observations (see [1] for details) but the mass window between 10^{17} and 10^{22} g remains unconstrained.

2 Hawking evaporation

S. Hawking demonstrated [2], by applying quantum field theory effects to BHs, that a radiation could be released outside their event horizon, reducing then the BH mass. This radiation spectrum follows an almost blackbody distribution and for photons, it writes:

$$\frac{d^2N}{dEdt} = \frac{1}{2\pi}\frac{\Gamma(E, M_{\rm PBH})}{e^{E/T_{\rm PBH}} - 1} \tag{1}$$

where $T_{\rm PBH} = M_P^2/(8\pi M_{\rm PBH})$ with M_P being the Planck mass and $\Gamma(E, M_{\rm PBH})$ is the greybody factor. This effect is not significant for stellar BHs but only for the lightest PBHs. However, it can be shown that PBHs formed with $M_{\rm PBH} < 5 \times 10^{15}$ g should already have evaporated by now. The PBH photon flux is calculated as follows:

$$\frac{d\Phi}{dE}(l,b) = \frac{f_{\rm PBH}}{4\pi M_{\rm PBH}}\frac{d^2N}{dEdt}\int_{\rm l.o.s.} ds\,\rho(r(s,l,b))\,. \tag{2}$$

where $\rho(r(s,l,b))$ is the DM density profile and s, l, b are the line of sight (l.o.s.), galactic longitude, and galactic latitude, respectively. In the following, the DM density profile is assumed to be a NFW profile [3]. For $M_{\rm PBH} = 10^{16-18}$ g, the PBH signal peaks at MeV energies and can be looked for in Galactic hard X-ray diffuse emission.

3 Galactic diffuse MeV emission

3.1 The MeV gap

Despite its major scientific interest, the Galactic diffuse MeV emission is poorly studied. The continuum sensitivity of high-energy photon telescopes and detectors worsen by several orders of magnitude around MeV energies and the only measurement of the Galactic diffuse MeV emission was performed by the Compton Telescope (COMPTEL) more than 20 years ago [4]. Bounds on $f_{\rm PBH}$ have been set using COMPTEL data [5], but more recent data from SPI, the SPectrometer on INTEGRAL, can also be used.

3.2 SPI data analysis method

At high energies, focusing photons becomes a hard exercise. Wolters telescopes, such as XMM-Newton, can be used up to keV energies, but for MeV and GeV photons, telescopes look like particle detectors. SPI is an instrument with a coded-mask made of transparent pieces that let the photons come in and opaque pieces absorbing them. On the germanium detector, located below the coded-mask, each source emitting photons casts a coded shadow, but the detected photons are not directionally tagged and this impacts the data analysis.

One analysis method is to assume spatial models for the photon emission, simulate what would SPI see through the coded-mask and compare the results to the SPI raw data, *i.e.* numbers of photons per observation, per energy and per detector. The best model is chosen with a maximum likelihood fit performed in each energy bin independently, which finally allows one to construct the total spectrum.

3.3 Galactic diffuse MeV spectrum from 16 years of SPI data

The major contribution to the diffuse MeV photons is known to come from Inverse Compton scattering (ICS) and we made a systematic study of this model component. The ICS spatial templates were created with GALPROP using one of the best-fit model *galdef* files from Fermi-LAT analyses [6]. Several alternative ICS models were created from our baseline model, such as a model with a thick halo or a model with a single cosmic-ray diffusion index. Additional templates, for unresolved sources, nuclear lines, and positronium annihilation, were included in the model. The significant instrumental background was also carefully studied.

Using the method previously described, we extracted the Galactic 0.5-8 MeV diffuse emission from 16 years of SPI data [7] from a $\sim 95^\circ \times 95^\circ$ region of interest. Our measurement (orange points in Figure 2 left) is in agreement with the old COMPTEL measurement (green points) but with a better signal to noise ratio (magenta points). Adding a NFW spatial template to the model led to no detection of a DM-origin signal. Therefore, upper limits can be put on $f_{\rm PBH}$.

4 Constraints on PBH DM

We used the newly extracted SPI spectrum to put constraints on PBH DM. We performed a Monte Carlo Markov chain spectral fit to adjust the various spectral parameters of the model component, namely the parameters $f_{\rm PBH}$ and $M_{\rm PBH}$ of the PBH spectrum, given by equation 1. The other model components are fitted with power-laws and Gaussians. We put the sampled values of $f_{\rm PBH}$ and $M_{\rm PBH}$ on a scatter plot and defined the 95% confidence level upper limit on $f_{\rm PBH}$ such as 95% of the samples falls below the limit for every mass bin.

Figure 2 right shows various constraints on PBH DM from the literature (colored broken lines) in the $10^{16} - 10^{18}$ g mass range, together with the bound we derived (solid black). We set the strongest constraints on PBH DM for $M_{\rm PBH} \gtrsim 10^{17}$ g and up to $M_{\rm PBH} \sim 4 \times 10^{17}$ g [8]. Above this value, PBH can still account for all the DM.

5 Conclusion

The DM is a major component of our Universe but its nature still hides its secret. PBH are non-particle DM candidates, and are by far the heaviest ones. These hypothetical objects could be detected for example, by observing their microlensing effects or by gravitational wave detectors, but the current non-detections allow one to set bound the abundance of PBH,

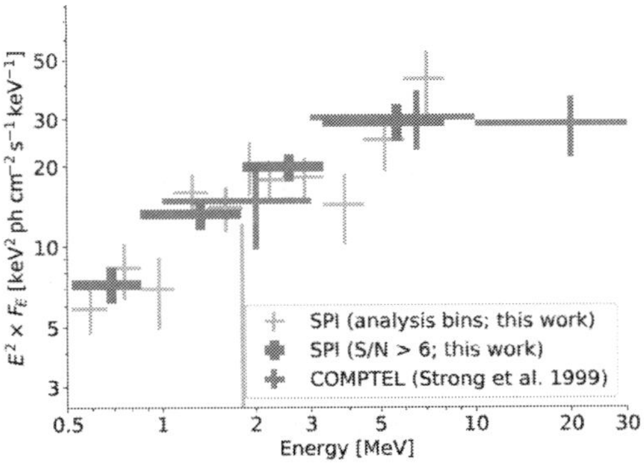

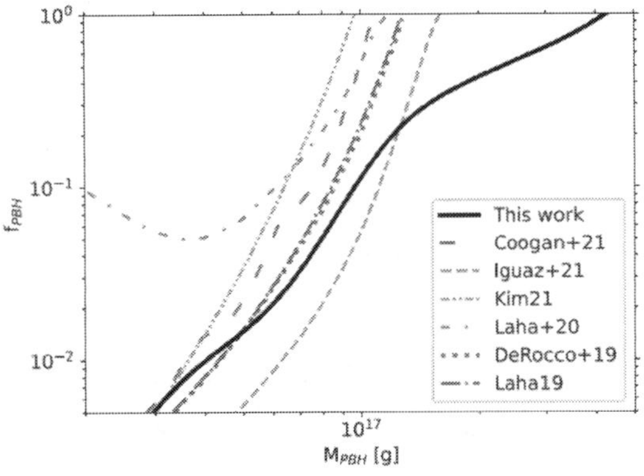

Figure 2. *Left:* Galactic diffuse MeV spectrum from 16 years of SPI data (orange) and from COMPTEL data (green). The SPI spectrum can be rebinned to obtain a signal to noise ratio larger than the COMPTEL measurement (magenta). *Right:* Upper limits on f_{PBH} as a function of M_{PBH} from the literature (colored broken lines) and from our analysis (black solid line). We showed that PBHs cannots saturate the DM cosmological abundance below $M_{\mathrm{PBH}} = 4 \times 10^{17}$ g.

i.e. on the fraction of DM in the form of PBHs. Light PBHs should be evaporating today in hard X-ray photons. We first extracted the Galactic diffuse MeV spectrum from 16 years of SPI data and then used it to put the strongest constraints on PBH DM in the asteroid mass range. We considered a simple model of non rotating PBH with a monochromatic mass distribution, and relaxing these constraints is expected to produce more stringent bounds.

A large PBH mass interval remains unconstrained and PBH could still saturate the DM cosmological abundance. Alternatively, our MeV spectrum can also be used to constrain decaying DM models, in which the DM candidates are axion-like particles or sterile neutrinos, as shown in [9].

References

[1] Green and Kavanagh, Journal of Physics G: Nuclear and Particle Physics **48**, 29 (2021)
[2] Hawking, Monthly Notices of the Royal Astronomical Society **152**, 75 (1971)
[3] Navarro, Frenk *et al.*, Astrophysical Journal **490**, 493 (1997)
[4] Strong, Bloemen *et al.*, Astrophysical Letters and Communications **39**, 209 (1999)
[5] Coogan, Morrison *et al.*, Physical Review Letters **126**, 171101 (2021)
[6] Ackermann *et al.*, Astrophysical Journal **750**, 35 (2012)
[7] Siegert, Berteaud *et al.*, Astronomy & Astrophysics **660**, 11 (2022)
[8] Berteaud, Calore *et al.*, Physical Review D **106**, 023030 (2022)
[9] Calore, Dekker *et al.*, arXiv hep-ph 2209.06299 (2022)

Scalable solutions for the Control Unit of the KM3NeT DAQ system

Cristiano Bozza[1*] for the KM3NeT Collaboration

[1]University of Salerno and INFN Gruppo Collegato di Salerno, Dipartimento di Fisica, Via Giovanni Paolo II 132, 84084 Fisciano, Italy

Abstract. The neutrino telescopes of KM3NeT are being incrementally expanded, and will reach their final size in the coming years. New versions of optical modules running new versions of firmware and new instrumentation for calibration are being introduced in the originally repetitive lattice. The inner architecture and data flow of the Control Unit of the KM3NeT telescopes is described, with information about computational and architectural complexity. The current goal is to control two full blocks of the KM3NeT/ARCA detector, i.e. 4370 CLBs and 128340 photomultipliers for 230 detection units, with a single mid-range commercial server machine. The system is designed with software protections and fault tolerance for hardware failure.

1 Introduction

The KM3NeT Collaboration is incrementally building and operating two neutrino telescopes in the Mediterranean Sea: ARCA[†], located off the shore of Sicily is devoted to neutrino astronomy ([1]); ORCA[‡], off the French shore is optimised for the study of neutrino oscillations ([2], [3], [4]). Both are water Cherenkov detectors with 3D lattices of *DOMs* (Digital Optical Modules) [5], each hosting 31 *PMTs* (photomultipliers) plus a compass, tiltmeter and humidity and pressure sensors, all controlled by a *CLB* (Central Logic Board); DOMs are grouped in DUs (*Detection Units*), which are vertical strings, few hundred meters tall, anchored to the seabed, of 18 regularly spaced DOMs. The spacing of DOMs is wider for ARCA than for ORCA, reflecting the different energy ranges of the neutrinos they are targeted at (up to PeV scale for ARCA, GeV scale for ORCA). In full configuration, ARCA will consist of two *blocks* of 115 DUs each, with in total 4140 DOMs, 128340 PMTs and 230 DU-base modules, instrumenting more than 1 km^3 of water; ORCA will have 115 DUs, 2070 DOMs, 64170 PMTs and 115 DU-base modules, covering 0.007 km^3 of water. Sea currents change the shape of the detectors; the position of detector elements is measured by an acoustic system including piezo elements in the DOMs, LED beacons and external hydrophones. A Calibration Unit and an Instrumentation Unit monitor water properties. Optical data from the PMTs and acoustic data from the piezos and

* Corresponding author: cbozza@unisa.it

†Astroparticle Research with Cosmics in the Abyss

‡Oscillation Research with Cosmics in the Abyss

hydrophones are transmitted to the control station on shore and processed by a computing farm, whereas data from other instruments are used for monitoring and recorded in a remote relational Database. At the time of the RICAP 2022 conference, 19 DUs were already taking data in ARCA and 10 in ORCA. At a given time KM3NeT has indeed more than two detectors in operation: each DU in the test benches at the end of the integration and before deployment is handled as an independent detector, managed by the same system.

2 The Control Unit

With the telescopes in their final configurations, about 400,000 parameters (HV, temperature, power, humidity, compass, tiltmeter, etc.) need to be controlled every 60 seconds in ARCA and 200,000 in ORCA. Some are input parameters (e.g. HV), others would raise alarms (e.g. humidity increase) if out of bounds. In addition, the *CU* (Control Unit) [6] checks the status of the *TriDAS* (Trigger and Data Acquisition System) and adjusts it according to the current task (calibration or physical data taking). DOM CLBs run two flavours of firmware and operate according to a state machine; DU-base module CLBs also have two variants of firmware for power control; a similar state machine also defines the behaviour of the TriDAS processes.

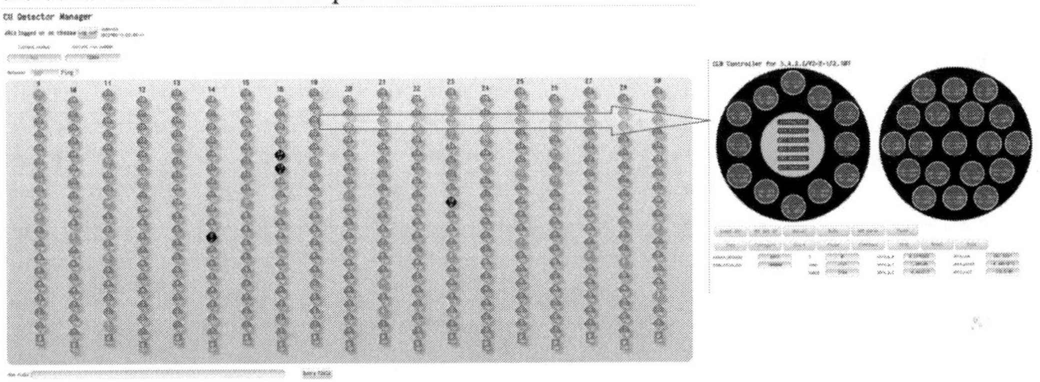

Fig. 1. Screenshot of the Detector Manager graphical user interface, showing the overview of the status of the full detector and a detailed view of one DOM.

The CU is a suite of fault-tolerant, partially independent software services. It has an overall state controller, the *MCP* (Master Control Program). The *DM* (Detector Manager) (Fig. 1) translates this into the set of actions needed to consistently govern each CLB, sets the operating parameters of PMTs (such as voltage and threshold) according to the *runsetup* defined on the MCP and reads slow monitoring and book-keeping data, staging them into a temporary area. The *TM* (TriDAS Manager) ensures that the *DataQueue* processes prepare data for the *Optical Data Filters* to perform online triggering, sending the results to a *DataWriter* for rolling storage in the local site before transmission to final storage; *Acoustic Data Filters* write their acoustic signal triggers directly to temporary storage. Finally, the *LAP* (Local Authentication Provider) provides user identification, dynamic role management of the machines, self-diagnosis and automatic load balancing in case one or more servers experience hardware problems. The *DPM* (Dynamic Provisioning Management) function of the LAP monitors local CU and TriDAS processes, preventing unwanted excess CPU or memory usage and restarting stopped processes in case of need. In case of degraded power, such as can happen if a server fails, LAPs of the other servers recognise the failure and reallocate the processes to the remaining machines. A *DBI* (Data

Base Interface) uploads datafiles from the local staging area to the remote Database and acts as a local cache to ensure continuity of data taking in case the Internet link to the remote Database is down. The most CPU-intensive task are handled by the DM. Currently in ARCA with 19 DUs there are 76 incoming UDP datagrams/s from the CLBs, each 175 B/s/CLB while the outbound UDP packets towards CLBs are as small as 19 B/s/CLB. The CPU load due only to monitoring tasks is 1.39 core on an Intel Xeon E5-2640@2.00 GHz. This results in 18 MB of monitoring data every 10 minutes, i.e. about 1 TB/year.

3 Performances

In its full configuration, the ARCA detector will have more than 4000 CLBs needing supervision and communication with the DM, which is the most critical component of the CU from the point of view of performances. A lightweight UDP-based messaging protocol has been developed for this application. Data must frequently be sorted by time or source, via optimised sorting methods. Writing continuously to disk or DB is inefficient, and all CU processes have an internal buffer to optimise writing tasks in bunches of 32 MB or 10 minutes. The monitoring information also needs to be presented in a GUI (*Graphical User Interface*), with hundreds of parameters represented in JSON and transported over HTTP to each subscribing client, the graphical rendering being left to the client's browser. Once the list of data to be extracted is defined, a direct-access path is created and no further sorting and searching occurs. A robust design must prevent thread starvation (i.e. thread pools exhausted or growing beyond the number of available cores) and deadlocks (both internal, due to data consistency requirements and external, due to cross-process logical error loops). In order to choose a safe attitude, multiple sockets are allocated with properly sized buffers and they are shared among CLBs in fixed groups (automatically defined); there is one readout thread per socket, which distributes data to CLB controller entities in the DM software, each with its own queue. This architecture cancels any chance of thread race. All data logging uses multiple memory slots to avoid thread races; logs are sorted at flush time, when data are copied to disk. Monitoring data from CLBs are processed by threads that are allocated as specified by configuration parameters: worker threads visit the CLB controller entities in a *round-robin* fashion, to log data in their queues and perform the needed reactions and adjustments. As a result of all this care taken in DM design, measurements done in regular data acquisition days in ARCA with 18 DUs show (Fig. 2) that the CPU load scales almost linearly and not faster than *N log N* (*N* being the number of DUs). Extrapolating these figures to a full *block* of 115 DUs, 4 CPU cores on the currently used server should suffice; a scalable solution with multiple DMs can be set up if needed.

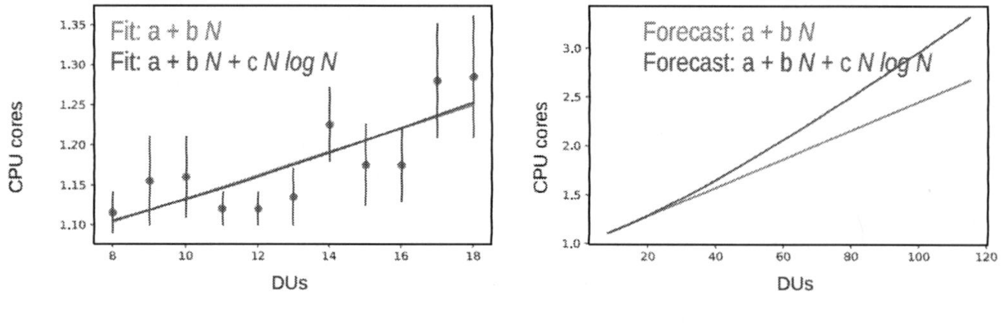

Fig. 2. CPU load (ARCA 18 DUs) on Intel Xeon E5-2640 v2 @ 2.00GHz (left) and extrapolation to full ARCA *block* of 115 DUs.

4 Overload protections

CU services are unlikely to cause CPU or memory overloads in normal conditions. On the other hand, TriDAS processes require normally a significant amount of resources and are already now spread over several servers. Protections are set up against runaway CPU or memory consumption. DataFilters and DataWriters are single-thread applications, so each one cannot saturate more than one core. This allows for intrinsic protections, by avoiding instantiating too many processes. DataFilter memory buffers are statically allocated. On the other hand, DataQueue memory buffers tend to grow if there is insufficient computing power (which may happen in case of hardware failure of one or more DataFilter servers). In this case, the resident LAP of each server watches for overall memory consumption and stops/restarts all local processes if the available memory falls below a tunable limit. In case DataWriter memory buffers grow too much, the resident LAP takes similar action.

5 Flexibility, code development and maintenance

The CU drives very different entities, because of the timespan of construction of the KM3NeT detectors. CLBs have two different variants and run two different versions of firmware. CLBs in base modules host a separate component for power functions that can have different version of firmware. Additional devices that need to be controlled by the CU are the Calibration Unit and the Instrumentation Unit, and more can appear in the future. The codebase must be flexible enough to allow evolution. The codebase is entirely in C# running on the Mono CLR and has only the Oracle Data Provider as external dependency to connect to the remote Database. Binary images are tagged and self-checked on startup, the fingerprint being added to all data. Loops, filters and sort operations are implemented in functional-programming fashion by means of LINQ, which makes the code expressive and readable. Behaviours are described with action tables. The source code for each device controller is kept in a single file. Event reactions are implemented as well-identified actions. Interprocess calls run on an HTTP library and also the graphical interface uses pure HTTP and Javascript. Static type checking and unit tests are used to actively prevent mistakes.

6 Conclusions

The Control Unit software for the KM3NeT Data Acquisition has been presented. It is a mature software suite, but evolving to support new versions of firmware for existing devices, new devices and adapting to changing scenarios. The code is produced to be clear and easy to maintain, with a mission to prevent, mitigate or eliminate risks of data loss. Performances are well under control and one commercial server should manage two *blocks* of 115 DUs. The architecture is fault tolerant and allows multiple resources standing by.

References

1. S. Aiello et al. (KM3NeT Collab.), Eur. Phys. J. C, 10.1140/epjc/s10052-021-09187-5
2. S. Aiello et al. (KM3NeT Collab.), JHEP, 10.1007/JHEP03(2022)055
3. S. Aiello et al. (KM3NeT Collab.), Eur. Phys. J. C, 10.1140/epjc/s10052-021-09893-0
4. S. Aiello et al. (KM3NeT Collab.), JHEP, 10.1007/JHEP10(2021)180
5. S. Aiello et al. (KM3NeT Collab.), JINST, 10.1088/1748-0221/17/07/P07038
6. S. Aiello et al. (KM3NeT Collab.), Comp. Phys. Comm., 10.1016/j.cpc.2020.107433

Investigating the time properties of an improved 3" Hamamatsu photomultiplier for the KM3NeT Neutrino Telescope

Andreino Simonelli[1,*], *Carlos Maximiliano* Mollo[1], and *Pasquale* Migliozzi[1],
on behalf of the KM3NeT collaboration

[1]INFN sezione di Napoli

Abstract. The Hamamatsu R14374-02 3-inch photomultiplier tube is an improved version of the previous R12199-02 model. It will be used for the completion of the KM3NeT neutrino telescope. Five hundred photomultipliers have been characterized for dark count rate, timing spread, and spurious pulses with a dedicated dark box apparatus.The results are compared with the model R12199 used in the first phase of construction of KM3NeT.

1 Introduction

KM3NeT is a research infrastructure composed of a network of deep-sea neutrino detectors located in the Mediterranean Sea [1–3]. The large number of involved photodetectors for th KM3NeT project raised the need for a detailed characterization of the PMTs. This is essential for the performance of the entire experiment. Both detectors share a common technology i.e a modular structure of multiple detection units that compose the matrix-like structure of both detectors. Each DU can be seen as a string of digital optical modules (DOMs) attached at a fixed distance between each other via a string kept almost vertical by a buoy. The DOM [7] is a 17-inch pressure resistant glass sphere with 31 3-inch PMTs inside. The time requirements for the KM3NeT PMTs were formerly reported in [5] and we report them again here in Table 1 for ease of reading. The Hamamatsu R12199-02 PMT was the photodetector chosen for the first phase of the KM3NeT detector. It is a 3-inch-diameter hemispherical PMT with 10 dynodes and standard bialkali metal photocatode. In a second phase Hamamatsu released an improved version of the tube named R14374. The scope of this paper is to present the measurements performed on the latter PMT and to compare some of them with those reported in [5],in order to determine compliance with the technical requirements of the experiment and to present the technical improvements.

*email: andreino.simonelli@na.infn.it

Photocathode diameter	> 72 mm
Nominal Voltage for gain 3×10^6	900÷1300 V
Peak-to-Valley ratio	> 2.0
Transit Time Spread (FWHM)	< 5 ns
Dark count rate (0.3 s.p.e. threshold, at 20 °C)	2000 c.p.s. max
Prepulses between −60 ns and −10 ns	1.5% max
Delayed pulses between 15 ns and 60 ns	5.5% max
Late afterpulses between 100 ns and 10 μs	15% max

Table 1: Requirements for the main characteristics of PMTs to be used in the KM3NeT detectors (s.p.e. is for single photoelectrons; c.p.s. is for counts per second).

Figure 1: The dark box apparatus and two trays of R14374 PMTs ready for test

The dedicated studies on critical parameters done for the former PMT model were replicated with the new model..

2 Characterising the time properties of the Hamamatsu R14374 vs R12199 photomultipliers

The time characteristics of single photo electron creation and detection as well as the intrinsic dark noise of a PMT are of fundamental importance for photon counting application. The measurements were conducted with a set of 500 PMTs equipped with the digital base used for the commissioning of the KM3NeT experiment.

 https://doi.org/10.1051/epjconf/202328008002

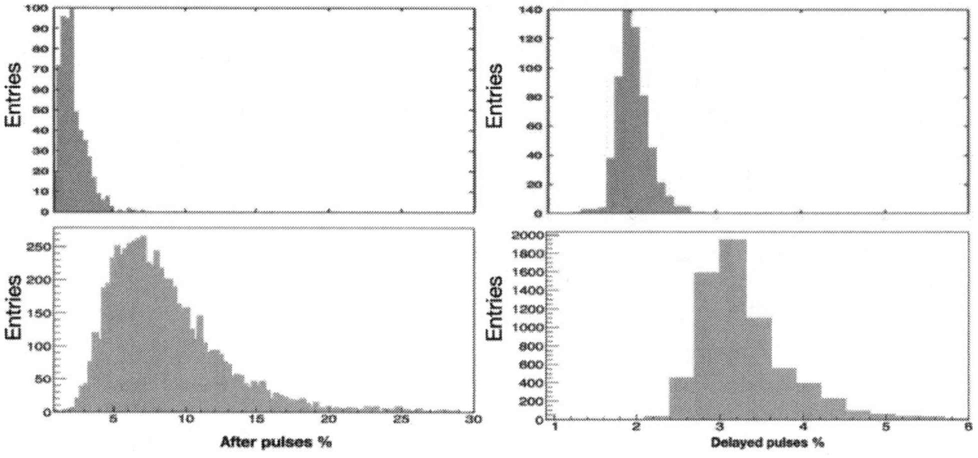

Figure 2: Measurements of the rate of delayed pulses and after pulses for a set of 500 PMTs R14374 (top row) and, for comparison, for the R12199 model as reported in[5].

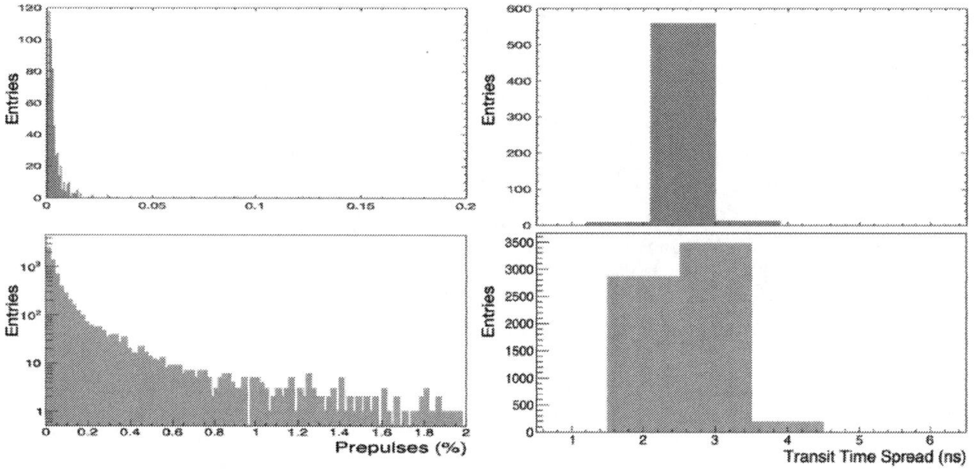

Figure 3: Measurements of the rate of pre pulses and transit time spread for a set of 500 PMTs R14374 (top row) and, for comparison, for the R12199 model as reported in[5].

The measurements were performed with the Dark Box described in [8]. The dark box replicates in a laboratory environment the structure of a downsized detector string composed of two digital optical modules. The dark box can host 62 PMT tubes per measurement and provides in a semi automatic way a report for every PMT in terms of dark noise counts, tuned high voltage for a specific gain, spurious pulses response properties: prepulses, afterpulses and delayed pulses. As a source, a continuously tunable repetition rate diode-pumped picosecond laser is used. The output of the laser is fed into a one to 62 optical fiber splitter with an intensity splitting accuracy below one percent. The time duration of a measurement is roughly one day. The results of the analysis produced by the dark-box apparatus are gathered and summarized statistically. This specific apparatus was previously used by [5] to perform measurements on a set of 5000 measurements of the previous R12199 model by Hamamatsu. The comparison in the analysis of delayed and afterpulses is reported in Figure 2. Figure 3 shows the results and the comparison of the old measurements on the bottom row compared to the new ones on top row for transit time spread and prepulses.

3 Conclusions

The measurements performed on a set of 500 PMTs and the comparison with the results reported in [5] reveal a large difference between the two models. The distributions of both afterpulses and delayed pulses are almost non overlapping in favour of the new model and also the dispersion of the distributions is narrower. The prepulses are reduced by more than one order on magnitude and the transit time spread is smaller and less scattered. Further research and a deeper characterization of the PMT are ongoing, but the results obtained already shows that the new model of 3" inch PMT that is already in use will improve the overall characteristics of the performance of the KM3NeT detectors.

References

[1] S. Adrián-Martínez *et al.* [KM3NeT Collaboration], "Letter of Intent for KM3NeT2.0", J. Phys. G **43** (2016) no.8, 084001; doi:10.1088/0954-3899/43/8/084001.

[2] S. Adrián-Martínez et al. [KM3NeT Collaboration], "Deep sea tests of a prototype of the KM3NeT digital optical module", Eur. Phys. J. C **74** (2014) no. 9, 3056; doi:10.1140/epjc/s10052-015-3868-9.

[3] S. Adrián-Martínez *et al.* [KM3NeT Collaboration], "The prototype detection unit of the KM3NeT detector", Eur. Phys. J. C **76** (2016) no.2, 54; doi:10.1140/epjc/s10052-014-3056-3.

[4] B. Herold, O. Kalekin , J. Reubelt, "PMT characterization for the KM3NeT project", Nucl. Instr. & Meth. in Phys. R. A **639** (2011) 70; doi:10.1016/j.nima.2010.09.018.

[5] Aiello, S., et al. "characterization of the Hamamatsu photomultipliers for the KM3NeT Neutrino Telescope." Journal of Instrumentation 13.05 (2018): P05035.

[6] P. Timmer, E. Heine and H. Peek, "Very low power, high voltage base for a photo multiplier tube for the KM3NeT deep sea neutrino telescope," JINST **5** (2010) C12049; doi:10.1088/1748-0221/5/12/C12049.

[7] S. Aiello et al [KM3NeT Collaboration], The KM3NeT multi-PMT optical module' 2022 JINST 17 P07038, DOI 10.1088/1748-0221/17/07/P07038

[8] C. M. Mollo *et al.*, "A new instrument for high statistics measurement of photomultiplier characteristics", JINST **11** (2016) no.08, T08002; doi:10.1088/1748-0221/11/08/T08002.

[9] S. Adrián-Martínez *et al.* [KM3NeT Collaboration], "A method to stabilise the performance of negatively fed KM3NeT photomultipliers", JINST **11** (2016) no.12, P12014; doi:10.1088/1748-0221/11/12/P12014.

[10] G. F. Knoll, Radiation detection and measurements, John Wiley Sons, Inc. (1999).

[11] B.K. Lubsandorzhiev et al., "Photo-electron backscattering in vacuum phototubes", Nucl. Instrum. Meth. A **567** (2006) 12; doi:10.1016/j.nima.2006.05.047.

[12] Hamamatsu Photonics K.K. Editorial Committee, "PHOTOMULTIPLIER TUBES Basics and Applications ", THIRD EDITION (Edition 3a).

The multi-PMT optical module of KM3NeT

Immacolata Carmen Rea[1,*] on behalf of the *KM3NeT Collaboration*

[1]INFN Sezione di Napoli

Abstract. The KM3NeT Collaboration has developed a multi-PMT optical solution for construction of the two deep-sea ARCA and ORCA telescopes. This approach has several advantages compared to the single-PMT optical modules which have been traditionally used in neutrino telescopes, including a larger photocathode area which is equivalent to three 10-inch photomultipliers. In addition, the multi-PMT approach allows for a high resolution, good positioning and timing calibration. The integration of the optical modules follows a strict protocol and a standardised procedure. It takes place in parallel at eight different integration sites in the labs of the collaboration. In this way, it is possible to reach a significant production rate of more than 100 well-qualified modules per month. In this proceeding, we present details of the technology of the KM3NeT optical module and the integration process.

1 Introduction

The extremely low neutrino cross-section and low high-energy neutrino fluxes motivated the scientific community to instrument cubic kilometers of water and ice with arrays of light detectors around the world. These light detectors, called "optical modules", are typically distributed in vertical structures and coupled to electro/optical cables. Such structures have a typical length of several hundreds of meters: they can be distributed according to specific geometries, optimized for the targeted energy region of the neutrino telescope. Moreover, the optical modules can host one or more Photomultiplier Tubes (PMTs), which can have different photocathode areas.

2 KM3NeT telescope design

The KM3NeT experiment is currently under construction at two different sites in the Mediterranean Sea:

- **KM3NeT/ARCA** (Astroparticle Research with Cosmics in the Abyss), 90 km offshore Portopalo di Capo Passero (Sicily) at 3500 m depth, aiming at the detection of HE cosmic neutrino sources ($E_\nu \sim$ GeV-PeV).
- **KM3NeT/ORCA** (Oscillation Research with Cosmics in the Abyss), 40 km offshore Toulon (France) at 2500 m depth, aiming at the study of the neutrino oscillations ($E_\nu \sim$ MeV-GeV).

*e-mail: imma.rea@na.infn.it

The two infrastructures share the same technology with different geometries, each optimized for its scientific goal. For ARCA are foreseen 230 Detection Units (DU), divided into two building blocks, to instrument 1 km^3 of water; for ORCA are foreseen 115 DUs (only one building block) that will instrument seven megatonnes of water.
To detect an adequate neutrino flux and to exploit the 70 m absorption length and the 100 m scattering length (in water, at 440 nm), sparse instrumentation is sufficient. For example, each 800 m string of ARCA hosts only 18 digital optical modules (DOMs).

3 The KM3NeT Digital Optical Module

The digital optical module of KM3NeT has an innovative design: it consists of a 17" diameter high-pressure resistant glass sphere that hosts a dense packing of 31 PMTs and calibration devices (for positioning and timing), electronics for power, readout, data acquisition, and transmission, fig.1 [1]. The segmented photocathode area of about 1200 cm^2 in each sphere

[a] [b]

Figure 1. Left: Picture of a multi-PMT digital optical module. **Right:** The components mounted inside a module.

increases the sensitivity of each DOM for the incoming direction of the detected photons: this leads to a broad angular coverage and good photon counting performance. Moreover, the impact of a PMT failure on the performance of the telescope is lowered as the module can still be operated efficiently with fewer PMTs.
The PMTs adopted for the KM3NeT optical module are the Hamamatsu R12199-02 (for the first optical modules) and R14374 [2], which has an improved timing performance (lower TTS). The PMT design is based on a convex bialkali photocathode with a 3" diameter and a 10-stage dynode structure. At an angle of 45° around the head of the PMT, a collection ring made of polished metal provides a 92% reflectance for photons in the 375-500 nm wavelength range [3]. In this way, the acceptance is increased by 20-40%. A light collection ring is shown in fig.2.

The calibration devices mounted inside the DOM, which are used to monitor its orientation with respect to the telescope coordinate system, are a compass and an accelerometer. Their data are sent to shore and converted into the orientation coordinates "Pitch", "Yaw", and "Roll" of the optical module. In addition, an acoustic piezo sensor glued inside the glass of the lower hemisphere constantly monitors the position of the module compared to a long baseline of acoustic emitters anchored to the seabed at known positions. This information is

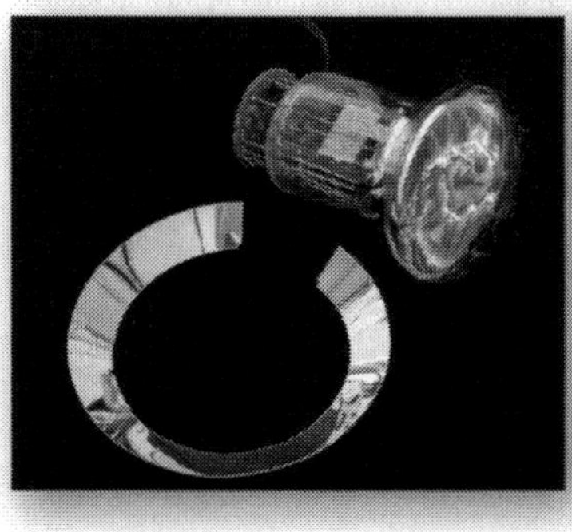

Figure 2. A photomultiplier tube with its light collection ring.

updated every minute and stored in the database.

Every DOM is equipped with a 470 nm LED pulsing circuit referred to as a "nanobeacon", that emits fast light pulses for timing calibration of neighboring DOMs. The intensity, timing, and pulse frequency of the LED can be controlled remotely from the onshore control station during dedicated timing calibration runs: the aim is to measure the propagation time of light from one module to the adjacent ones.

For the KM3NeT experiment, the PMT photon-signal measurement is reduced to two values only: the photon arrival time and the time-over-threshold (ToT). In this way, the required bandwidth for data transmission is minimized but it is still possible to deliver sufficient information. Then, all data collected offshore are digitized and sent without reduction to shore, where a farm of processors, with data trigger and selection algorithms, reduces the data volume and filters the signals from the background. Afterwards, the observables needed for physics analysis are identified. This approach requires that the electronics clocks of the DOMs are synchronized to sub-nanosecond precision. Moreover, the nanosecond accuracy on the measurement of the arrival times of the photons helps to filter out the background due to ^{40}K decay and bioluminescence.

4 Production model

In its final configuration, KM3NeT will consist of more than 6000 DOMs. In order to build such a large number of optical modules in a reasonably short amount of time, the production has been distributed among over eight integration sites inside the Collaboration. For this, the DOM integration procedure has been optimized for protection against errors and maximization of the production rate: as of today, the KM3NeT DOM integration sites are able to guarantee a baseline production rate of 100 fully qualified DOMs per month.
To build an optical module, the procedure foresees that the two glass hemispheres are filled bottom-up. The integration follows a strict protocol, and is composed of several steps that can be mainly listed as follows: 1) gluing a cooling mechanical structure in the top hemisphere and the acoustic piezo in the bottom hemisphere, 2) mounting the electronics and a penetrator, through which the DOM can be connected to a backbone cable of the DU, followed by a helium leak test and optical fiber splicing, 3) installation of the PMTs and light collection rings into the PMT support structures, 4) functional test, 5) pouring of a transparent two-component silicone gel that ensures optical contact between the PMTs and

the glass, 6) DOM closure, 7) final acceptance test. A dedicated software called "KM3DIA" (KM3 DOM Integration Assistant) guides the integrators through the right sequence of operations, logs all relevant information, and registers all DOM details. In this way, the history of all the components can be tracked in a central database, including functional and acceptance test results.
This organization ensures significant advantages, but it also requires a very high level of quality control and quality assurance for procurement of components and logistics.

5 Performance

Given the typical rate of the deep-sea environment, to minimize the effect of ageing (thus maximizing the lifetime of the PMTs), a low nominal gain of $3*10^6$ has been chosen. This corresponds to a ToT for a single photoelectron signal of 26.4 ns for a PMT read with a KM3NeT base.
At such gain value, the ToT response of the PMTs detecting multiple photons has been studied in lab measurements and in situ. The shape of the distribution of the ToT values due to single photoelectron signals can be described with an analytic model that takes into account as free parameters the *gain* and the *gain spread*. Following the model, the PMT gain can be monitored in situ and, if needed, the High Voltage (HV) can be adjusted.

Thanks to the multi-PMT design, it is possible, even with a single DOM, to distinguish the signal due to atmospheric muons from the background induced by ^{40}K and bioluminescence, exploiting multi-PMT coincidences, as shown in fig.3.

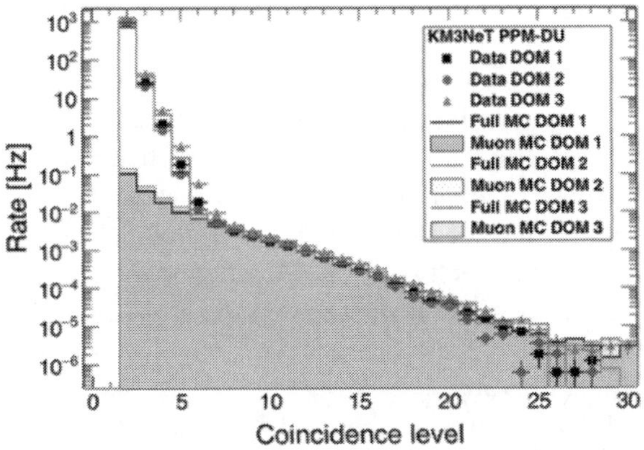

Figure 3. Rate of coincident photon detection as a function of the number of coincident PMTs for three different optical modules. This study has been made on the optical modules of the Pre-Production Module DU (PPM-DU) of KM3NeT

References

[1] The KM3NeT multi-PMT optical module, S. Aiello et al, JINST **17**,P07038 (2022)
[2] Characterisation of the Hamamatsu photomultipliers for the KM3NeT Neutrino Telescope, The KM3NeT Collaboration, S. Aiello, et al. JINST **13**,P05035 (2018)
[3] Expansion cone for the 3-inch PMTs of the KM3NeT optical modules, The KM3NeT Collaboration, S. Adrián-Martínez et al. JINST **8** T03006 (2013)

The KM3NeT data acquisition system - Status and evolution

Tommaso Chiarusi[1,2,*], *Emidio* Giorgio[2,**], and *Daniele* Zito[2,***] on behalf of the KM3NeT Collaboration

[1]INFN-Sezione di Bologna, Viale Berti-Pichat 6/2, 40127 Bologna, Italy
[2]INFN-Laboratori Nazionali del Sud, via S.Sofia 62, 95123 Catania, Italy

Abstract. In this contribution, we present the data acquisition system system of the KM3NeT neutrino telescopes, ARCA and ORCA, already operating while under construction at the bottom of the Mediterranean Sea. The DAQ system implements a modular and scalable hardware-triggerless streaming readout of the optical modules in the telescopes. Handling a raw incoming throughput up to hundreds of Gbps, it is integrated with the online alert system of the KM3NeT Multimessenger program. The system will evolve according to a new design, currently under validation, which will grant the extension of the telescope to the aimed cubic kilometer scale.

1 Introduction

The data acquisition (DAQ) system is an integral part of the KM3NeT infrastructure [2], responsible for collecting and storing data from the undersea Cherenkov detectors. Currently under construction in two different abyssal sites of the Mediterranean Sea, the ARCA and ORCA KM3NeT telescopes feature a common DAQ design which, although with different implementations, is designed to be operational since the first stages, scaling with the detector size. It includes also the control and monitoring systems of the off-shore installation as well as the on-shore infrastructure, which is made of the time-synchronization GPS reference, the networking fabrics, and the computing and storage resources. Although running with two independent DAQ installations, the two telescopes data taking are combined at higher level by the KM3NeT online Multimessenger system [3] In this contribution all the key elements of the DAQ system for the already running detectors are reviewed, together with the ancillary procedures to deploy, upgrade and maintain it. Eventually, the future DAQ paradigm, the *Full White Rabbit scenario* based on the White Rabbit paradigm [4] and currently under review, is presented.

2 The KM3NeT detector and the data streams

The KM3NeT detection elements such as the Digital Optical Modules (DOMs), the Detection Units (DU) and the Detection Unit-Base Modules (DU-BM), as well as the DOM and

*e-mail: tommaso.chiarusi@bo.infn.it
**e-mail: emidio.giorgio@lns.infn.it
***e-mail: daniele.zito@lns.infn.it

DU-BM readout electronics (the Central Logic Board - CLB) are reported in details in these proceedings ([1]) and in [2]. DOMs and DU-BMs are the nodes of the submarine branch of a global ethernet network infrastructure, which grants the communication between the detector and the related shore station, and provides the DOMs and DU-BMs with the distribution of the absolute time with a sub-nanosecond precision by means of the White Rabbit protocol. The data taking of each DOM is independent to the others, without any local trigger mechanism implemented in the DOM electronics. For this reason data from the DOM photomultipliers (PMTs) and the acoustic sensors are acquired according to a continuous streaming readout (triggerless) mode. Recorded data are packed in continuous sequences of *optical* and *acoustic* frames, each referring to a 100 ms time-slice duration. The digitisation of the PMT pulses after each photon-electron conversion leads to a 6-byte *hit* information, which includes the PMT identification, a hit-time reference and the duration of the pulse above a certain threshold. With PMT thresholds set at 0.5 photoelectrons, average ~6 kHz and ~8 kHz hit rates are measured for ARCA and ORCA, respectively, mostly due to the ^{40}K decays and bioluminescence contamination. A 20 kHz acquisition veto per PMT is set to avoid the contamination by a too large optical background. The average measured DOM optical throughput is 12 Mbps,while the acoustic streaming accounts for roughly the half. Globally, the throughput from such fast acquisition data streams is ~350 Mbps per DU. The DAQ system is designed around a conservative throughput value which is at least a factor three larger than the measuerd one, targeting to O(100) Gbps per building block. Few precents of the global throughput is due to the ancillary instrumentations, sampled at a minor rate (from 0.1 to 1 Hz), the slow-control data exchanges and the White Rabbit stream.

3 The DAQ system in the *Broadcast* scenario

The *Broadcast* scenario implies an asymmetric connection topology: a single downlink channel from the shore-station; one direct uplink per CLB back to on-shore resources ([5]). Such design is already active in the current installation of ARCA (21 DUs) and ORCA (14 DUs). It will serve both detectors up to the completion of the first deployment stage, i.e. 32 DUs and 48 DUs, respectively. The evolution of the connection design for the next deployment stages is reported in Section 7. For the asymmetry of the Broadcast scenario, the firmware of the White Rabbit switches (WRSs) used in the shore station and of the CLB has been customised, diverting from the the CERN Standard. Only the DU-BMs are in a true Master-Slave relation with the WRSs on shore, while the DOMs piggyback the incoming White Rabbit stream for synchronising, sending back to shore only the rest of the data streams to a fabric of standard (not WR) switches. Indeed, such standard high-throughput switches (various DELL® models) are configured with a dedicated implementation of the Software Defined Networking technology ([8]).

4 The DAQ processing chain

On-shore, the incoming optical and acoustics data streams are routed to the so called *Trigger and Data Acquisition System* (TriDAS), a multi-stage processing chain realised with C++ software applications. The first stage is the aggregation layer. A few concurrent processes called *DataQueue* collect the incoming optical and acoustic streams sent by a sector of the telescope, fragmented in various UDP datagrams, and reconstruct the frames recorded by the CLBs. The DataQueues route the frames downstream via TCP links. A valid reference for UDP and TCP protocols is [7]. The second stage is the filtering/processing layer, made of a number of concurrent processes called Optical and Acoustic *DataFilters*. Each Optical DataFilter receives from the DataQueues the entire set of CLB frames recorded during a given timeslice. Different processes simultaneously work on data recorded during different timeslices. Each Optical DataFilter implements various causality-trigger algorithms, requiring precise time-space constraints for the recoreded hits, and quickly identifies possible tracks of neutrino induced muons or shower-like events. An additional trigger algorithm selects data that will be later scrutinised as candidate neutrino events from supernovae[9]. The reduction of the incoming stream after the triggers is smaller than $1 : 10^4$. On each Optical DataFilter, a circular buffer stores raw data which can be dumped and analysed off-line after external alerts. The trigger algorithms used in the Optical Data Filters are implemented in a KM3NeT C++ programming framework, called Jpp [10], which is also used for off-line analyses and Monte Carlo production, ensuring identical processing for both real data and simulations. The Acoustic Data filters reconstruct the detection-time, also referred to as time of arrival (TOA), of the sound emitted by various submarine beacons, each corresponding to a given sound-waveform, by means of cross-correlation algorithms. Computed for each DOM and DU-BM, the TOAs are at the base of the off-line detector positioning system. The last stage is the data-writing and storage layer. By means of a custom data-dispatcher of Jpp which operates as middleware, each Optical DataFilter asynchronously transmits the selected data to a single collecting process, the DataWriter, which writes the incoming stream into ROOT files [11] on local storage systems.The reconstructed TOA are written to disk directly by the Acoustic DataFilters, in a dedicated binary format. Two copies of the same recorded files are secured in two different remote repositories: CNAF-INFN, in Italy, and CC-Lyon, in France.

5 Management of the DAQ context

The number of the DataQueue, Optical and Acoustic DataFilter processes necessary to guarantee a smooth data-taking depends both on the global throughput of the detector and the complexities of the filtering algorithms. The larger the data-load to process, the more resources are allocated. This is achieved by distributing replicas of the same process on different high performing servers (>32 cores at 2GHz each, > 128 GB RAM), which are interconnected via a 25 GbE networking infrastructure. This approach is called the MUltiple SErvers with DIfferent Processes (MUSEDIP) scenario, whose modularity by desing allows for scaling with the situation complexity. The TriDAS processes are orchestrated together with the detector by another software process, the Control Unit, whose latest evolution is described in details in these proceedings ([6]). The DAQ framework, beyond the ARCA and ORCA cases, is also exploited in other contexts, such as the DOM, DU-BM and DU integration and validation setups. TriDAS and Control Unit processes are deployed via a dedicated framework called AIACE (Automatic Installation And Configuration Environment): based on the ANSIBLE technology, AIACE allows to define a set of environments, one for each DAQ installation, and then to proceede from scratch with the installation and configuration of all the devices with the requested applications. All software is deployed as Docker ([12]) images, activated in due containers by AIACE. This approach is extremely efficient for easily implementing and maintaining the MUSEDIP scenario in all the DAQ contextes.

6 The Monitoring and the Multimessenger processing branches

The selected events, with some complementary information, are mirrored to both the monitoring system and the alert system for the Multi-Messenger program. The monitoring system is a collection of web appplications written in Python language which shows the activity status of the detector parts, plots of the the trigger algorithms efficiency, the synchronisation status and the quality of the data taking. The Multi-messenger alert system is a further branch of online contex, separately active in both ARCA and ORCA shore stations, performing a more accurate analysis of the triggered events, assigning them a quality score. High score events are sent to a central processing facility, presently running in the ORCA shore station, where a final analysis step is performed by searching for signals from supernovae or other transient phenomena. This is done by combining the selected data with external alerts as well as releasing alerts via the GCN [13], the SNEWS 2.0 [14] and other systems. In order to send alerts with the shortest delay, the full DAQ processing plus the online seraches are finalised within 15 seconds from the reception of the first raw data on-shore.

7 Evolution of the DAQ system

The design goal of two building blocks for ARCA cannot be achieved within the Broadcast scenario. In fact, in order to guarantee all the additional uplinks for the CLBs yet to be deployed, it would be necessary a number of optical fibres larger than what is available with the already implemented sea-floor network. For this reason, new aggregation layers will be implemented directly in the DUs: a pair of White Rabbit switches will be placed into each DU-BM, called *Wet*-White Rabbit switches (WWRS), after the redefinition of the form factor of the switch backplane to fit the DU-BM space. The two WWRS will connect all the CLBs of one DU, allowing also *cold* redundancies, and provide 2x 1Gbps ethernet uplinks to shore. Such an innovative approach, called *Full White Rabbit* scenario, implies various design changes. The main ones affecting the DAQ concern both the detector and the shore station: the new DOM-CLBs will become pure White Rabbit Slave units, with a new dedicated firmware. On-shore, standard White Rabbit switches will be used for implementing the front-end switch fabric (called *Dry* WRS). Additional advantages of this new design will be the reduction of connections on-shore by at least one order of magnitude with respect to the Broadcast scenario; the direct support by the international White Rabbit community (no customisation is foreseen in the White Rabbit firmware); a simplified time-calibration procedure for the CLBs embedded in the standard White Rabbit protocol. Despite the differences, Broadcast and Full White Rabbit parts of the detector will contribute together to the data taking, and their data streams will be combined at the TriDAS and Control Unit levels. At present, the Full White Rabbit approach is under production-readiness review. The new DUs following this design are expected to be produced by the end of 2023.

References

[1] Simone Biagi in these Proceedings
[2] S. Adrián-Martínez et al., J. of Physics G: Nucl. and Part. Phys, **43**, 084001 (2016)
[3] W. Assal et al., J. of Instrumentations, **16**, C09034 (2021)
[4] White Rabbit. `http://www.ohwr.org/projects/white-rabbit`
[5] S. Aiello et al., arXiv:2210.13328, accepted for publication on JINST
[6] S. Aiello et al., Computer Physics Comminications, **256**, 107433 (2020)
[7] L. Dostàlek, A. Kebalova, *Understanding TCP/IP* (Packt Publishing, 2006)
[8] T. Chiarusi, E. Giorgio, EPJ Web of Conferences, **207**, 06009 (2019)
[9] S. Aiello et al., Eur. Phys. J. C, **82**, 317 (2022)
[10] The JPP framework, `https://sftp.km3net.de/documentation/Jpp/v17.2.0/`
[11] The ROOT system, `https://root.cern.ch/`
[12] The Docker system, `https://www.docker.com/`
[13] The Gamma-ray Coordinates Network, `https://gcn.gsfc.nasa.gov/`
[14] The SuperNovae Early Warining System, `https://snews.bnl.gov/`

https://doi.org/10.1051/epjconf/202328008005

Virgo and the gravitational interferometry

Maddalena Mantovani[1,*] for the Virgo Collaboration

[1]European Gravitational Observatory (EGO), I-56021 Cascina, Pisa, Italy

Abstract. The Advanced Virgo detector is a long scale enhanced Michelson interferometer placed in Italy, close to Pisa, with the aim of detecting gravitational waves from astronomical sources. The Advanced Virgo interferometer has detected, together with the LIGO interferometers located in USA [1], an impressive collection of gravitational waves emissions in the last observation runs O2 and O3. During the last observation run (O3), which lasted about one year of data taking from April 2019 to March 2020, were detected about 80 events. When the run ended, the detector has been upgraded toward the Advanced Virgo + configuration, in order to enhance its sensitivity, which it is currently in the commissioning phase. After an introduction on gravitational wave detection, the paper will focus on giving an overview of the evolution of the Virgo antenna in the past decades.

1 Introduction and gravitational wave detection principle

The gravitational waves (GWs) have been predicted in 1916 by Albert Einstein [2] as quadrupole emissions creating ripples in the space-time due to extremely violent astrophysics events (e.g. supernova explosions, collisions of neutron stars and black holes, etc...). Due to the weak interaction between the gravitational waves and the matter, their direct detection occurred only one century after the prediction.
The technological advancements allowed the LIGO-Virgo collaboration to detect the GW emission from two coalescing black holes on the 14th of September 2015 (GW150914) [3].
As it has been mentioned before, the GW modifies the space-time metric in a quadrupole way, which means that modifies the distance measured between free-falling masses in a differential way for orthogonal directions. For this reason the most promising detection principle results to be the use of a Michelson interferometer, detecting the GW passage as a deviation from the interference condition [4]. The change of differential length due to a GW signal of strain amplitude h measured by a Michelson interferometer having orthogonal arms with length L is equal to::

$$\Delta L = \frac{hL}{2} \tag{1}$$

In order to have an idea of the order of magnitude of the observable displacement, it can be considered that for an arm length of 1 km the GW having a strain amplitude of 10^{-22} the generated displacement is very small of the order of $\Delta L = 10^{-19}$ m.
For this reason the optical configuration of the GW detector is much more complex than a simple Michelson interferometer.

*e-mail: μ maddalena.mantovani@ego-gw.it

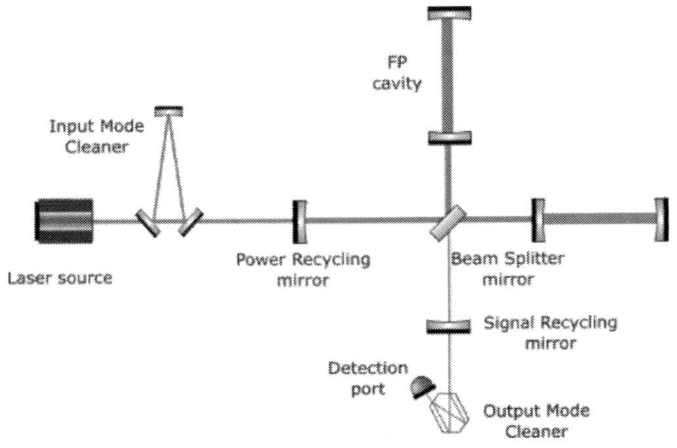

Figure 1. Schematic drawing of the Virgo detector in which the main optical components are shown as: the Fabry-Perot cavities (FP cavities), 3km long orthogonal arms, the Power Recycling mirror (PR), the Signal recycling mirror (SR), the Beam Splitter mirror (BS) which splits the incoming beam generated by the laser source in the two cavities and defines the interference condition at the detection port. Moreover two additional cavities are implemented to clean and stabilize the input and output beam, the Input Mode Cleaner (IMC) and the Output Mode Cleaner (OMC) respectively.

Additional cavities are added to enhance the detector response to GW signals, as shown in Figure 1:

- the Fabry-Perot cavities in the long arms. The addition of a high reflecting mirror in the long arm allows to have the laser beam to bounce between the two mirrors increasing the laser power stored in the cavity, to reduce the Shot-Noise limit of the sensitivity, and to increase the optical path length of the arm.
- the Power Recycling mirror (PR). The addition of the PR between the laser and the interferometer increases the recycled light to improve the Shot-Noise limit.
- the Signal Recycling mirror (SR). The SR mirror is placed between the Beam-Splitter mirror (BS) and the detection port to recycle the GW signal enhancing the detector response.

The addition of all these optical resonators allowed to improve the detector sensitivity up to the first detection of GW emission in 2015, but on the other hand the increase of complexity of the detector has strongly increased the difficulties in its commissioning. The main reason is that the detector provides the best sensitivity only if all the optical cavities are actively controlled to maintain precise resonant conditions [1][5][6].
Moreover, the sensitivity of the detector is not limited only by the Quantum-Noise (Radiation pressure and Shot Noises), but also other fundamental noises contribute to the sensitivity, such as the thermal noise [2] and the seismic noise. The technological enhancements pushed these limits towards unprecedented levels.

2 The Virgo detector and its evolution

The Virgo experiment has been started in 1997 thanks to a French-Italian collaboration [7]. The Virgo collaboration has grown in time, being currently a collaboration formed by 800

[1] In order to have an idea of the accuracy requirements of the control on the position of the main optical component it can be said that the differential displacement of the two long arms has to be lower than 10^{-16} m.

[2] The thermal noise is due to the thermal excitation of the mechanical modes of the test masses and the suspensions, resulting in a spurious signal at the detection port

	Virgo	Advanced Virgo	Advanced Virgo+
Input power [W]	17	25	33
FP *Finesse*	50	450	450
Arm cavity geometry	Plano-Concave	quasi-confocal	quasi-confocal
Power Recycling gain	50	38	38
Signal Recycling			✓
Fused silica fibers	for VSR3	✓	✓
Scientific joint runs	VSR1, VSR2, VSR3	O2, O3	spring 2023
Maximum BNS range [Mpc]	17	60	
GW detection		✓	

Table 1. Summary table to show the main parameters of the Virgo detectors.

members, 534 authors, 136 institutions from 15 countries. The installation and commissioning of the Virgo detector was completed in June 2006 [8], the main optical parameters are listed in the first column of Table 1 1 [3]. For the first time three scientific runs had been carried out in coincidence with the LIGO detectors, namely VSR1, from March to October 2007, and VSR2, from July to December 2009 and the last VSR3, from June to October 2011. The major change between the VSR2 configuration and the VSR3 is the modification of the bottom part of the payload by substituting the steel wires hanging the test masses with fused silica fibers, in order to kill the suspension thermal noise.
The mean BNS range [4] achieved by the Virgo detector in the VSR3 run was about 17 Mpc. Unfortunately the sensitivity reached in the VSR* runs was not enough to allow a GW detection.
After the end of the VSR3 observing run, the detector has been upgraded toward the Advanced Virgo experiment [11]. The long arm cavity geometry has been strongly changed, from plano-concave to quasi-confocal configuration, to reduce the test mass thermal noise. Moreover to improve the Shot-noise and to increase the optical path length the arm finesse has been strongly increased of a factor 9 respect to the Virgo configuration, see second column of Table.
The strong increase of the power stored in the arm cavities, and the marginality of the recycling cavity stability, has lead to strong effort in developing and implementing a Thermal Compensation system (TCS) to cope with thermal transients and defects [12]. The Advanced Virgo experiment allowed to detect on the 14th August 2017 the first triple coincidence LIGO/Virgo, BH-BH merging GW170814. Only few days later, on the 17th of August, the first detection of binary neutron star system occurred. The detection has lead to an intense observing campaign in the electromagnetic follow-up [13]. The O2 and O3 runs contributed to detect more than 80 GW events.
The O3 observing run was concluded in March 2020, at the beginning of the pandemic emergency, the scientific data taking period was stopped to start the installation of the Advanced Virgo+ experiment.
The main differences between the Advanced Virgo+ and the Advanced Virgo detector [14] are (see third column of Table 1):

- addition of the Signal recycling mirror, see Figure 1, to improve the detector response

[3]The *Finesse* of an Fabry-Perot cavity is a parameter which defines how narrow the resonance is, and it is proportional to the cavity gain which is the ratio between the stored power and the input power [10].

[4]The BNS range is defined as the distance at which a bynary system of two neutrons stars (having 1.4 solar masses each) could be detected with a signal to noise ratio of 8.

- increase of the input power, to reduce the quantum noise at high frequency (shot-noise)
- addition of frequency dependent squeezing [15], to improve the broadband quantum noise

The installation of the Advanced Virgo+ has been concluded in December 2020, after that the detector entered in the commissioning phase to put in operation the aforementioned upgrades. The interferometer has been commissioned for almost two years, reaching a stable operability of the detector. Now it is currently in the phase of optimization and noise hunting to achieve the design sensitivity to start the O4 joint run, in the spring of 2023.
The main detector parameters for the three Virgo configurations are summarized in Table 1.

3 Conclusions

The Virgo detector, together with the LIGO ones, has evolved in time reaching unprecedented sensitivities allowed to open a new window on the universe by detecting a impressive large amount of GW emissions.
The detector is currently in the Advanced Virgo+ configuration which is meant to start the scientific data taking, in a joint run, in the next spring.

References

[1] Aasi J et al. (LIGO Scientific Collaboration) 2015 Class. Quant. Grav. 32 074001
[2] Einstein, A.; Die Grundlage der allgemeinen Relativitatstheorie, Ann. Phys. 49, 769-822, 1916.
[3] Abbott, R. et al. (LIGO Scientific Collaboration and Virgo Collaboration); Observation of Gravitational Waves from a Binary Black Hole Merger. Phys. Rev. Lett.,116, 061102, 2016.
[4] Nardecchia, I. Detecting Gravitational Waves with Advanced Virgo. Galaxies 2022, 10, 28. https://doi.org/10.3390/galaxies10010028
[5] Acernese, F.; et al. The Advanced Virgo longitudinal control system for the O2 observing run. Astropart. Phys. 2020, 116, 102386.
[6] Allocca, A., et al. "Interferometer Sensing and Control for the Advanced Virgo Experiment in the O3 Scientific Run". Galaxies 2020, 8, 85. https://doi.org/10.3390/galaxies8040085
[7] "VIRGO - Final Design Report" - https://dcc.ligo.org/T970234/public
[8] T. Accadia, et al. "Status of the Virgo project" Class. Quantum Grav. 28 (2011) 114002
[9] J. Abadie *et al.* VIRGO and LIGO Scientific collaboration, "Sensitivity Achieved by the LIGO and Virgo Gravitational Wave Detectors during LIGO's Sixth and Virgo's Second and Third Science Runs," [arXiv:1203.2674 [gr-qc]].
[10] A.E. Siegman, ", University Science Books, Mill Valley, 1986
[11] The Virgo collaboration, "Advanced Virgo Technical Design Report" https://tds.virgo-gw.eu/ql/?c=18777
[12] L Aiello et al 2019 J. Phys.: Conf. Ser. 1226 012019
[13] Abbott, B. P.; et al. (LIGO, Virgo and other collaborations) (October 2017). The Astrophysical Journal. 848 (2): L12. arXiv:1710.05833
[14] The Virgo Collaboration, 'Advanced Virgo Plus Phase I Design Report', Virgo-Technical Documentation System, Report No. VIR-0596A-19 (2019) https://tds.virgo-gw.eu/ql/?c=14430
[15] Jean-Pierre Zendri, et al. 'Advanced Virgo Squeezer Technical Design Report' LVK internal note, https://tds.virgo-gw.eu/ql/?c=11140, VIR-0411C-15

Overview of feebly interacting particle in indirect detection searches

Federica Giacchino[1,*]

[1]INFN Sezione Roma Tor Vergata and SSDC (ASI)

Abstract. Axion-like particles (ALPs) as feebly interacting particles (FIPs) candidate have generated a lot of interest in the recent years. Stars are good FIPs factories and consequently those particle can be detected through their interaction with the interstellar medium or decay in standard particles. Briefly, I will illustrate how high-energy astrophysics observations can be exploited to set constraints to ALPs model and to the ASTRO-MeV gap problem.

1 Introduction

What likely makes up the 84% of the matter content in the Universe is called Dark Matter (DM). All evidence of DM is purely gravitational and covers larger and larger scales. Anyway, its identification still remains an unsolved problem of the physics Beyond the Standard Model (BSM). Indeed no SM particles match the inferred DM characteristics. In the last years, feebly interacting particles (FIPs) [1] have been representing an interesting alternative to the so-called WIMP, namely Weakly Interacting Massive Particles. The FIPs are relatively light particles (sub-GeV masses), with extremely suppressed interactions with SM bosons or fermions. They have sound theoretical motivations of particle physics and in particular they could be the manifestation of a complicated secluded sector, called *dark sector* connected with the visible sector by a "mediator" known as *portal*. The portal is representative of low-dimensional operators according to the nature of the new particle: i.e Axion-Like Particle (ALP).

2 Axion-Like Particles

An ALP is a light pseudoscalar particle arising from the spontaneous breaking of an approximate global symmetry, a pseudo-Goldstone boson. According to the quasi-exact global symmetry, it identifies a particular BSM particle, i.e. the QCD axion [2]. A peculiar characteristic is that the mass and the coupling constant of ALPs are independent parameters and, for this reason, they scan a wide mass range by direct and indirect detection techniques. The ALP plays the role of DM itself or as a mediator through the dark sector. An ALP is coupled to ordinary matter and radiation and the Lagrangians involving these fields enjoy an approximate shift symmetry under which $L_a \to a + const.$, which implies that their interactions has to proceed via derivative terms.

*e-mail: federica.giacchino@roma2.infn.it

2.1 Production in Astrophysical environment

The stars are very good factories of FIPs. For the ALPs, according to the interaction, the production processes inside the star could be: Primakoff $\gamma + Ze \rightarrow a + Ze$, Compton Scattering $\gamma + e \rightarrow e + a$, Electron Bremsstrahlung $e + Ze \rightarrow e + Ze + a$ and nucleon Bremsstrahlung $N_1 + N_2 \rightarrow N_3 + N_4 + a$. Each process is dominant in a particular stage of the star lifetime and ALP mass range [3].

After production, the FIPs can free stream and escape from the core inducing novel channel of energy-loss and impacting the stellar evolution or producing spectral signatures in cosmic background as in gamma-ray. Experiment as Fermi-LAT [4] are able to catch this kind of indirect detection signals.

According to the nature of the coupling, the signatures to search are different: *i)* for very light mass, conversion of ALPs into photons in the Galactic magnetic field can produce spectral irregularities in galactic and/or extragalactic source spectra, i.e. core collapse SNe [5] or diffusive gamma-ray background [6]; *ii)* for heavy mass, ALPs can be detected through their interaction with the interstellar medium or decay in SM particles inducing spectral features in the expected background [7] [8].

Unfortunately the MeV energy region, where many objects in the Universe emit, is never uncovered by any good performant instrument: COMPTEL reached a sensitivity at around $10^{-9} \frac{erg}{cm^2 \cdot sec}$, INTEGRAL hardly reaches the soft gamma-ray regime with an insufficient sensitivity, the FERMI pair vertex tracker telescope has an high energy threshold around 30 MeV and a bad point spread function below 100 MeV. Therefore the so-called ASTRO-MeV gap is empty by research. An intriguing and low-cost idea could be a crew of nano-satellites (1U=1 unit), i.e. COMCUBE [9], which detect energies from 100 keV to $\mathcal{O}(1)$ MeV, with very good instrument response functions (preliminarily the 4U COMCUBE sensitivity of $10^{-11} \frac{erg}{cm^2 \cdot sec}$ is obtained). The idea is to use COMCUBEs to detect observable fluxes of ALPs from pre-SN phase as Betelgeuse [10] and to scan the ALP parameter space in the MeV gap. The collaboration between Bari University Group and Roma Tor Vergata University is promoting that kind of analysis.

References

[1] P. Agrawal, *et al.* Eur. Phys. J. C **81** (2021) no.11, 1015
[2] R. D. Peccei and H. R. Quinn, Phys. Rev. Lett. **38** (1977), 1440-1443
[3] G. G. Raffelt, Lect. Notes Phys. **741** (2008), 51-71
[4] M. Meyer, M. Giannotti, A. Mirizzi, J. Conrad and M. A. Sánchez-Conde, Phys. Rev. Lett. **118** (2017) no.1, 011103
[5] M. Crnogorčević, R. Caputo, M. Meyer, N. Omodei and M. Gustafsson, Phys. Rev. D **104** (2021) no.10, 103001
[6] C. Eckner, F. Calore, P. Carenza, M. Giannotti, J. Jaeckel, F. Sivo and A. Mirizzi, PoS **ICRC2021** (2021), 543
[7] A. Caputo, H. T. Janka, G. Raffelt and E. Vitagliano, Phys. Rev. Lett. **128** (2022) no.22, 221103
[8] F. Calore, P. Carenza, M. Giannotti, J. Jaeckel, G. Lucente, L. Mastrototaro and A. Mirizzi, Phys. Rev. D **105** (2022) no.6, 063026
[9] A Laviron et al., *SF2A-2021: Proceedings of the Annual meeting of the French Society of Astronomy and Astrophysics. Eds.: A. Siebert* (2021) 105-108
[10] M. Xiao, K. M. Perez, M. Giannotti, O. Straniero, A. Mirizzi, B. W. Grefenstette, B. M. Roach and M. Nynka, Phys. Rev. Lett. **126** (2021) no.3, 031101

Sensitivity of KM3NeT/ARCA to a flux of nuclearites

Alice Păun[1,2,*], *Gabriela* Păvălaş[1], and *Vlad* Popa[1] *on behalf of the KM3NeT Collaboration*

[1]Institute of Space Science, RO-077125 Bucharest, Măgurele, Romania
[2]University of Bucharest – Faculty of Physics, RO-077125, Măgurele, Romania

Abstract. Over the past decades, theories have predicted the existence of heavy compact objects containing an extremely dense form of exotic matter named Strange Quark Matter (SQM). This type of hypothetical matter is composed of nearly equal quantities of up, down and strange quarks and is supposed to be the ground state of Quantum Chromodynamics. Nuclearites are the massive component of SQM particles. Some studies show that nuclearites heavier than 10^{13} GeV with velocities of approximately 250 km/s could reach the Earth and could be observed by neutrino telescopes. In this paper, Monte Carlo simulations are used in order to evaluate the preliminary results on the sensitivity of KM3NeT/ARCA to a flux of down-going nuclearites.

1 Introduction

Nuclearites, or Strange Quark Matter (SQM) aggregates, are hypothetical compact and massive particles introduced in 1984 [1, 2]. They are composed of three quark flavours (up, down and strange) [2] and their structure is similar to that of an atom [2]. The detection of SQM aggregates can be done through the blackbody radiation emitted in the visible spectrum along its path, that occurs at the elastic collision between nuclearites and atoms of the medium [2]. Due to this characteristic, nuclearites could be detected by neutrino telescopes. KM3NeT is a network of deep-sea neutrino telescopes located in the Mediterranean Sea, dedicated to the search for high-energy cosmic neutrinos and the study of neutrino properties. The KM3NeT detectors are currently under construction and they are already taking data with the first installed lines.

2 Analysis and results

For this study, an adapted Monte Carlo (MC) simulation code conceived for ANTARES detector was used. The code was modified to correspond to the new geometries, Digital Optical Modules (DOM) and PMT characteristics [3]. The analysis uses MC simulated nuclearite events for masses in the range $3 \cdot 10^{13} - 10^{17}$ GeV/c^2 for the full ARCA configuration. Several distributions were done, relevant for the expected nuclearite signal at the KM3NeT/ARCA depth and the corresponding test cuts were applied (Figure 1 - A, B, C), regarding the three discrimination variables: the triggered hits (at least two hits in a time interval on the same DOM), the snapshot hits (all hits in a time interval larger than the triggered event) and the snapshot duration (the time interval between the last and the first hit observed, with an extra amount of time added).

*e-mail: alice.paun@spacescience.ro

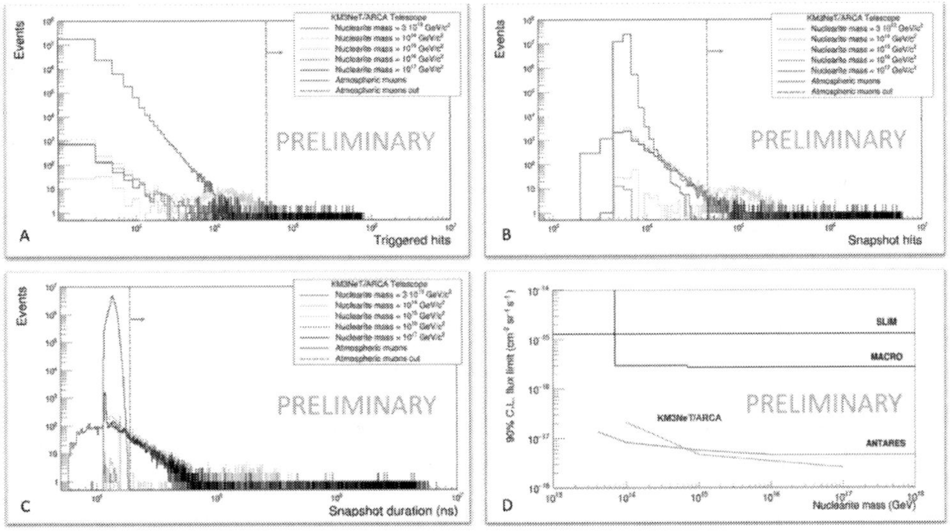

Figure 1. (A) Triggered hits, (B) Snapshot hits and (C) Snapshot duration distributions for nuclearites and atmospheric muons, with optimized cuts through MRF. (D) Preliminary sensitivity results of KM3NeT/ARCA detector to nuclearite signal compared to MACRO [4] and SLIM [5] upper limits and ANTARES [6] sensitivity for 839 days of 2009-2017 data.

The background due to ^{40}K decay is added in the filtering process and the simulated atmospheric muon component is removed by using optimized cuts. The acceptance and sensitivity of KM3NeT/ARCA to a flux of down-going nuclearites were computed, considering the snapshot duration criteria, and the sensitivity was compared to MACRO [4], SLIM [5] and ANTARES (839 days of 2009-2017) [6] results. The ARCA sensitivity at 90% C.L was computed by using the Feldman-Cousins prescription [7], considering nuclearite events with a Poisson distribution (Figure 1 - D).

3 Conclusions

Three variables were evaluated in order to find the best nuclearite selection criteria. The most promising variable was found to be the snapshot duration of the events. The preliminary sensitivity of KM3NeT/ARCA, obtained considering this criteria, was determined for a flux of massive down-going nuclearites at 90% C.L. This result is comparable to and could improve the ANTARES sensitivity obtained for 839 days of 2009-2017 data.

References

[1] E. Witten, Physical Review D **30**, 272-285 (1984)
[2] A. De Rujula, S. L. Glashow, Letters to Nature **312**, 734–737 (1984)
[3] A. Păun et al., JINST **16** (2021)
[4] M. Ambrosio et al., Eur. Phys. J. C **13**, 453–458 (2000)
[5] S. Cecchini, M. Cozzi, D. Di Ferdinando et al., Eur. Phys. J. C **57**, 525–533 (2008)
[6] Bouta et al., Journal of Instrumentation **16** (2021)
[7] G. J. Feldman, R. D. Cousins, Phys. Rev. D **57** (1998)

Tracking millisecond pulsars responsible for the Fermi GeV excess

Joanna Berteaud[1,2,*], *Francesca* Calore[1], and *Maïca* Clavel[2]

[1]LAPTh, CNRS, USMB, F-74940 Annecy, France
[2]Univ. Grenoble Alpes, CNRS, IPAG, F-38000 Grenoble, France

Abstract. More than 10 years ago, an excess of γ-ray photons coming from the Galactic center was discovered in the Fermi-LAT data. First attributed to dark matter, it has since been shown that it should have at least a partial stellar origin. One hypothesis is the presence of a population of millisecond pulsars (MSPs) confined in the Galactic bulge. We here present our recent progress in the selection of MSP candidates.

1 The Fermi GeV excess and the millisecond pulsar hypothesis

Millisecond pulsars (MSPs) are neutron stars with periods up to 30 ms. They have a magnetic field strength from 10^7 to 10^{11} G, which produces a beamed emission responsible for the pulsar lighthouse effect. MSPs are known to emit in radio, X- and γ-rays. They were proposed as an explanation for the Fermi GeV excess, a signal of γ-ray photons coming from the Galactic center region not reproduced by astrophysical models of diffuse γ-ray emission [1]. In particular, the spatial distribution of the photons making up this excess emission was found to correlate with old stars in the Galactic bulge [2, 3], and recent evidence for point sources below the telescope detection threshold supported this hypothesis [4].

In [5], we have developed a phenomenological model for X-ray MSPs in the Galactic bulge, whose γ-ray emission can fully explain the excess. We predict about 3×10^4 MSPs in the Galactic bulge. The X-ray flux for each MSP is computed from a gamma-to-X correlation we built from the sample of known MSPs. For more details we refer to [5]. Our conclusion was that a hundred MSPs could have been detected in past observations in our region of interest, $6^\circ \times 6^\circ$ around the Galactic center. We found more than 3000 non extended, non variable Chandra sources with a power-law-compatible spectrum in this same region. This demonstrated that the MSP explanation to the GeV excess was not excluded by X-ray data.

The Monte Carlo simulation used in [5] is now publicy available here. Our simulation includes MSP position (see figure 1 left), γ-ray luminosity, γ-to-X flux ratio and X-ray spectral index, which allowed us to compute their X-ray flux. The probability density functions we used were either found in the literature or computed them from published data.

2 Promising candidate selection and radio follow-up

Among the 3000 Chandra sources mentioned in section 1, we selected those with neither ultraviolet nor optical possible counterparts as promising MSP candidates [6]. We then defined two categories: 1) **compact objects** *i.e.* about 40 candidates with only a faint infrared

*e-mail: μberteaud@lapth.cnrs.fr

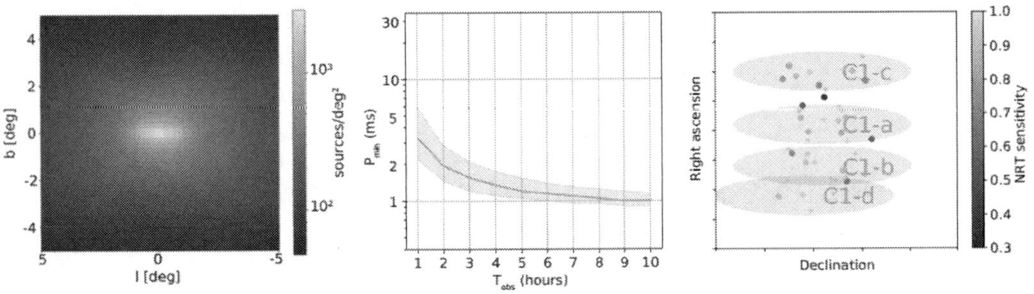

Figure 1. *Left:* Positions of MSPs in our simulation. *Middle:* Minimal detectable MSP period as a function of GBT observing time for one promising candidate. The line assumes a radio spectral index $\alpha = -1.9$ and the shaded area considers $-2.43 < \alpha < -1.37$. *Right:* NRT observation strategy. The orange ellipses show the NRT beam. Colored dots indicate the positions of the most interesting MSP candidates and the sensitivity of the telescope at this position. Gray crosses represent other candidates.

counterpart verifying $\log_{10}(F_X/F_{IR}) > 0.5$ [7] and 2) **radio sources** *i.e.* about 10 promising candidates with only a radio counterpart detected in unpublished VLA 1.4 GHz images. These promising targets motivated dedicated radio follow-up studies which are essential to detect pulsations.

We predicted the optimal radio observation time to undoubtedly confirm the MSP nature of some of our radio sources with Parkes and the GBT (see figure 1 middle) and were granted more than 55 hours of observations. We also obtained about 18 hours of radio observation with the NRT to observe two clusters of MSP candidates. The observation strategy was optimized to cover the maximum number of candidates in a minimal amount of time with a maximal sensitivity (see figure 1 right). Data reduction and analysis are currently on-going and our preliminary results are encouraging.

3 Conclusion and perspectives

By modeling the multi-wavelength MSP emission in X and gamma rays, we have defined a list of most promising X-ray bulge MSP candidates. Radio observations will conclusively adjudicate on their pulsar nature. Finding MSPs will be a huge step forward in understanding the GeV excess, but finding no pulsars will also be interesting. If our candidates are not pulsars, then what could they be? Finally, Einstein Telescope and CTA will help solve the excess mystery in the future, from a multi-messenger point of view [8, 9].

References

[1] Simona Murgia, Annu. Rev. Nucl. Part. Sci. **70**, 455-438 (2020)
[2] Richard Bartels *et al.*, Phys. Rev. Lett. **116**, 051102 (2016)
[3] Oscar Macias *et al.*, Nature Astronomy, **2**, 387 (2018)
[4] Francesca Calore *et al.*, Phys. Rev. Lett. **127**, 161102 (2021)
[5] Joanna Berteaud *et al.*, Phys. Rev. D **104**, 043007
[6] Joanna Berteaud *et al.*, PoS(ICRC2021)681 (2021)
[7] Dacheng Lin *et al.*, ApJ **756**, 18 (2012)
[8] Francesca Calore *et al.*, Phys. Rev. Lett. **122**, 081103 (2019)
[9] Oscar Macias *et al.*, MNRAS **506**, 1741-1760 (2022)

The contribution of unresolved sources on the gamma-ray spectrum of molecular clouds

Vittoria Vecchiotti[1,*], *Giada* Peron[2,**], *Giulia* Pagliaroli[3], and *Francesco Lorenzo* Villante[3,4]

[1]Gran Sasso Science Institute, viale Francesco Crispi, 67100 L'Aquila, Italy
[2]Astroparticule et Cosmologie, 10 Rue Alice Domon et Léonie Duquet, 75013 Paris, France
[3]INFN, Laboratori Nazionali del Gran Sasso, 67100 Assergi (AQ), Italy
[4]University of L'Aquila, Physics and Chemistry Department, 67100 L'Aquila, Italy

Abstract. Fermi-LAT observations of the large-scale diffuse gamma-ray emission unveiled that in some locations, towards the inner Galaxy, the spectrum of this component is harder than the local one, measured at the Earth position. If this signal is associated with the "truly" diffuse emission produced by the interaction of cosmic rays (CRs) with the interstellar medium (ISM) then its observed spectral features can be interpreted as indirect evidence of a CR spectral hardening toward the Galactic center. However, in order to correctly interpret the data, the contribution from unresolved sources has to be taken into account. Newly developed theoretical models showed that the cumulative flux produced by unresolved pulsar wind nebulae (PWNe) added to the "truly" diffuse emission significantly shapes the spectrum of the large-scale diffuse emission, challenging the CR spectral hardening hypothesis. In light of the recent results, we discuss the effect of unresolved PWNe on the observed spectra of the diffuse emission with a particular focus on molecular clouds (MCs). We analyze the influence of unresolved sources on clouds of different sizes and locations. Finally, we provide a prescription on how to choose the regions to target in order to have an unbiased determination of the "truly" diffuse emission.

1 Introduction

Galactic diffuse emission in the GeV energy band is mostly produced by the interaction of CR nuclei with the dense ISM. Observations of this radiation bring priceless information on the spatial and spectral distribution of CRs in the Galaxy far from the Solar System.

Gamma-ray astronomy has been extensively used to probe the CR spectrum across the Galaxy targeting both the large-scale diffuse emission (e.g. [1]) and MCs [2, 3]. These observations showed an enhanced gamma-ray emission and harder spectrum around ~ 4 kpc, compared to what is measured at Earth. The causes for this deviation are still unclear. This finding was interpreted as evidence of a progressive CR spectral hardening towards the Galactic center. Nevertheless, the inferred CR spectral index from known MCs shows a significant variation within the same ambient [2, 3]. These fluctuations could be related to the higher density of CR sources in the enhanced ring. The presence of nearby sources would significantly change the gamma-ray emissivity depending on the relative distance cloud/accelerator(s).

*e-mail: vittoria.vecchiotti@gssi.it
**e-mail: giada.peron@apc.in2p3.fr

Another possible explanation involves faint sources that produce a flux that does not overcome the detection threshold. These unresolved sources constitute a non-negligible component that adds to the "truly" diffuse emission, namely the one contributed by the bulk of CRs interacting with the ISM, and shapes the measured large-scale diffuse emission. In previous studies, their contribution was considered relevant only at very high energy ($\gtrsim$ 1 TeV) and was constrained to be less than 20% at GeV energies [1]. However, these results were obtained disregarding the contribution of PWNe. Recently, it was demonstrated that unresolved PWNe can shape the spectrum of the large-scale diffuse gamma-ray emission probed by Fermi-LAT in the GeV energy range and eventually explain the CR spectral hardening observed in the inner Galaxy [4].

In this work, we investigate the influence of such a component on the observed gamma-ray spectra from MCs. Being the unresolved source contribution subject to the observed angular size, we expect it to be less important when observing smaller regions. If compact clouds are unaffected by this additional diffuse signal, that would make them an unbiased barometer to probe the CR density in the Galaxy.

2 Methods and results

We calculated the expected diffuse contribution from MCs according to:

$$F_\gamma = N_{col}\Delta\Omega\,\xi_N \int dE_p \frac{d\sigma(E_p,E_\gamma)}{dE_\gamma} J_p(E_p,\bar{r}_{MC}) \propto \frac{M_5}{d^2_{kpc}}\xi_N \int dE_p \frac{d\sigma(E_p,E_\gamma)}{dE_\gamma} J_p(E_p,\bar{r}_{MC})$$

where N_{col} is the column density of the gas, $\Delta\Omega$ is the angular size covered by the MC, ξ_N accounts for the fraction of heavy nuclei in the CRs and in the ISM, $d\sigma/dE_\gamma$ is the proton-proton differential cross section and $J(E_p,\bar{r}_{MC})$ is the flux of protons at the location of the interaction; M_5 is the mass of the MC in units of 10^5 $M_\odot$ while d_{kpc} is the distance from us in kpc. In the following, we assumed both a CR spectrum that matches the local measurements, as modeled in [5], and a harder CR spectrum of index 2.5, normalized to the local measurements. In this analysis, we focus on the specific region centered at $(l,b) = (330,0)^\circ$ for different possible $\Delta\Omega$. This region has been chosen because it overlaps with the position of a well-known MC from the catalog of [6] and because, at this location, the contribution from unresolved sources (modeled as in [4]) is maximized and eventually emerge above the truly diffuse signal as a combined result of the source spatial distribution and the gas distribution. In the above estimate, the truly diffuse emission is calculated for our two hypotheses on the CR spectrum by assuming the target gas as traced by Planck opacity measurements at 353 Hz. The ratio between the unresolved and the truly diffuse components is shown in Figure 1 for the different spectral assumptions and as a function of energy. As expected, the influence of unresolved sources increases with energy and with the angular size. As clouds in the Galaxy, with the exception of local clouds, are $\lesssim 1^\circ$ extended (rightmost panel of Fig. 1), the flux of unresolved sources is expected to be less than 30% even where the unresolved contribution peaks. To test the above observation, we perform the Fermi-LAT analysis at the MC location $(l,b) = (333.4, 0.4)^\circ$ following the methods of [3] to extract the spectrum of the cloud and of the surrounding (10-deg wide) medium. In addition to the standard sources from the Fermi-LAT catalog, we include in the model the upper limit and the lower limit of the unresolved component, as derived by [4]. We extract the spectrum in the hypothesis of low, high, and no contribution of unresolved sources.

The results are shown in Fig. 2. In the case of the cloud, no variation of the spectral points is detected in the different cases, therefore, the contribution of unresolved sources seems not to affect the chosen cloud. Since it is located in a region where this contribution is maximized and has $\Delta\Omega \sim 1^\circ$ larger than the average angular size of other clouds, we expect

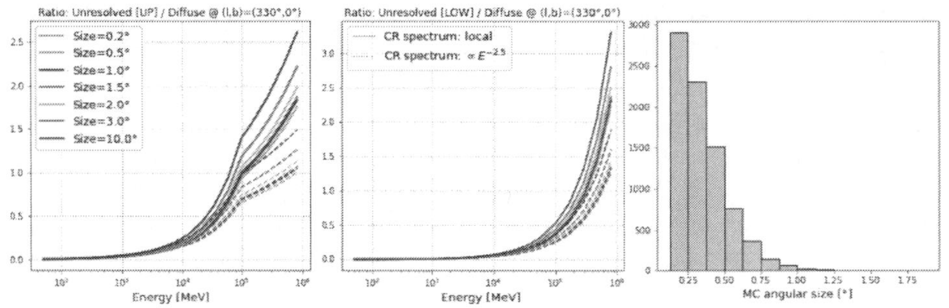

Figure 1. The ratio between the unresolved sources and the truly diffuse emission integrated over different areas. The right-most panel shows the distribution of extensions of MCs in the Milky Way [6]

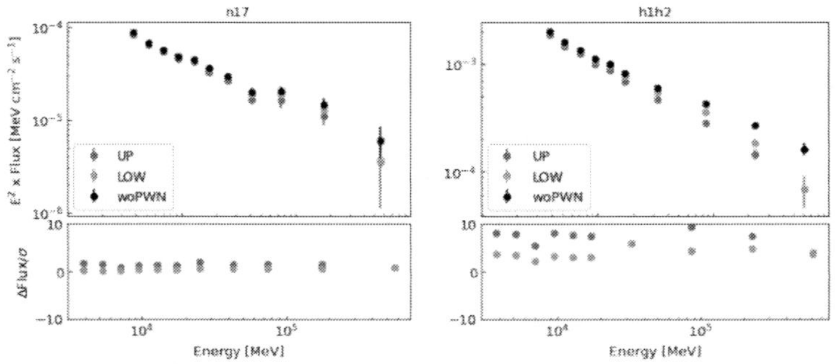

Figure 2. Spectral energy distribution (SED) of the cloud (left) and its surrounding medium (right) obtained including the lower (orange) and upper (blue) limit for the unresolved sources in the model. Black points show the SED obtained without including the unresolved contribution. In the lower panels we show the difference from the spectral points obtained with and without the contribution of unresolved sources in the lower- and upper-limit case, as a fraction of the statistical uncertainty.

that for $\Delta\Omega \leq 1^\circ$ the influence of unresolved component is negligible. This changes when we look at the surrounding area. Here the diffuse emission is modified by the presence of unresolved sources and it shows clearly a harder spectrum when the unresolved contribution is not included. More targets are under analysis to confirm this finding. This will allow us to localize regions in the plane where the truly diffuse emission can be extracted without contamination from unresolved sources and, therefore, constrain the propagation of CRs in the Disk.

Acknowledgements

The work of GP is partially supported by the research grant number 2017W4HA7S (MIUR).

References

[1] M. Pothast, D. Gaggero, E. Storm, C. Weniger, Journal of Cosmology and Astroparticle Physics **2018**, 45 (2018)

[2] F. Aharonian, G. Peron, R. Yang, S. Casanova, R. Zanin, Physical Review D **101** (2020)

[3] G. Peron, F. Aharonian, S. Casanova, R. Yang, R. Zanin, Astrophysical Journal Letters **907** (2021)

[4] V. Vecchiotti, G. Pagliaroli, F.L. Villante, Communications Physics **5** (2022)

[5] G. Peron, F. Aharonian, Astronomy & Astrophysics **659**, A57 (2022)

[6] M.A. Miville-Deschênes, N. Murray, E.J. Lee, The Astrophysical Journal **834**, 57 (2016)

Neutrino predictions from choked Gamma-Ray Bursts and comparison with the observed cosmic diffuse neutrino flux

Angela Zegarelli[1,2,3,*], *Michela* Fasano, *Silvia* Celli[1,2], *Dafne* Guetta[4], *Antonio* Capone[2,4], and *Irene* Di Palma[1,2]

[1]Dipartimento di Fisica dell'Università Sapienza, P.le Aldo Moro 2, I-00185 Rome, Italy
[2]Istituto Nazionale di Fisica Nucleare, Sezione di Roma, P.le Aldo Moro 2, I-00185 Rome, Italy
[3]Dipartimento di Fisica, Università di Roma Tor Vergata, Via della Ricerca Scientifica, 00133 Rome, Italy
[4]Department of Physics, Ariel University, Ariel, Israel

Abstract. The strong constraints from the Fermi-LAT data on the isotropic gamma-ray background suggest that the neutrinos observed by IceCube might possibly come from sources that are hidden to gamma-ray observations. A possibility emerged in recent years is that neutrinos may come from jets of collapsing massive stars which fail to break out of the stellar envelope, and for this reason they are known as choked jets, or choked Gamma-Ray Bursts (GRBs). We here show our predictions of neutrino flux and spectrum expected from these sources, focusing on Type II SNe, through detailed calculations of $p\gamma$ interactions and accounting for all the neutrino production channels and scattering angles. We provide predictions of expected event rates for ANTARES, IceCube, and the next generation neutrino telescope KM3NeT. We also compute the contribution of the choked GRB population to the diffuse astrophysical neutrino flux, thus providing constraints on the local rate of this source population as to reproduce the observed neutrino flux.

1 Introduction

Supernovae (SNe) explosions originated from core collapse of massive stars can be accompanied by a Gamma-Ray Burst (GRB)-like outflow that could be unable to break through the stellar envelope [1]. In such a situation, we talk about a *choked GRB*, with gamma-rays hidden to observations. Despite the lack of gamma-ray emission, choked GRBs are potentially emitters of high-energy neutrinos. For this reason, these sources have attracted much attention in the last years because of the possibility of explaining the all-flavor cosmic neutrino flux observed by IceCube [2] without incurring into inconsistencies with the isotropic diffuse gamma-ray emission from neutrino sources detected by the Fermi satellite [3].

2 Predicted neutrino fluxes from choked GRBs

In this contribution, we summarize the results of our work [4], that readers are invited to refer for details. We performed Monte Carlo simulations of $p\gamma$ interactions taking place in a choked jet, from which neutrinos arise, accounting for all the neutrino production channels

*e-mail: μangela.zegarelli@roma1.infn.it

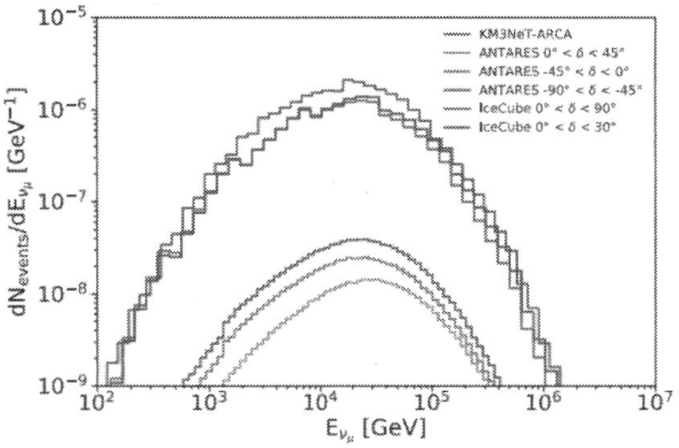

Figure 1: Number of events per energy bin induced in neutrino telescopes by ν_μ from an individual choked GRB.

Detector	Declination δ	N_{events}
ANTARES	$0° < \delta < 45°$	2×10^{-3}
	$-45° < \delta < 0°$	3×10^{-3}
	$-90° < \delta < -45°$	5×10^{-3}
KM3NeT/ARCA	Mean δ	2×10^{-1}
IceCube	$0° < \delta < 90°$	1×10^{-1}
IceCube	$0° < \delta < 30°$	2×10^{-1}

Table 1: Expected total number of events induced in neutrino telescopes by ν_μ from an individual choked GRB. The numbers in the table are obtained by integrating the curves in Fig. 1.

and scattering angles. Specifically, we computed the all-flavor time-integrated neutrino flux on Earth from type II SNe, by adopting the following benchmark values characterizing the jet features: Lorentz factor $\Gamma = 100$, jet lifetime $t_{jet} = 10^3$ s, and isotropic luminosity $L_{iso} = 10^{50}$ erg/s.

Our predictions of the number of ν_μ events expected from an individual choked GRB and detectable by ANTARES, IceCube, and the next generation neutrino telescope KM3NeT are reported in Fig. 1 and Tab. 1. We also computed the contribution of the choked GRB population to the diffuse astrophysical neutrino flux, providing constraints to the local rate of this source population as to reproduce the observed neutrino flux [5, 6]. We find that for GRB energies channeled into protons $E_{p,jet}$ between $10^{51} - 10^{53}$ erg, choked GRBs may contribute to the observed astrophysical neutrino flux, if their local rate is $R_0 = 80 - 1$ Gpc^{-3} yr^{-1}, respectively. If $E_{p,jet}$ is decreased, the local rate of choked GRBs has to increase in order to reproduce the IceCube data.

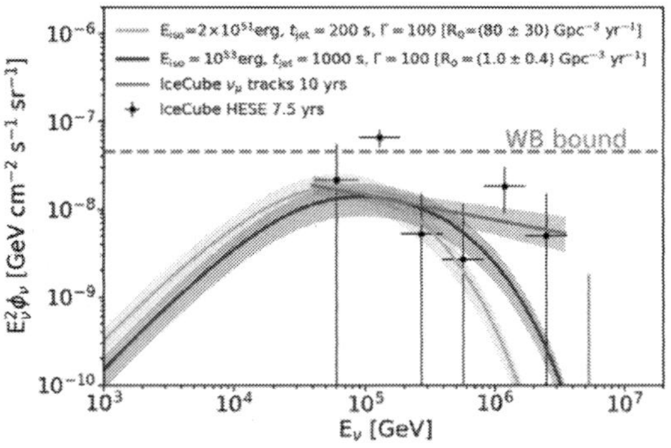

Figure 2: Diffuse neutrino spectra from choked GRBs expected on Earth, as compared with the IceCube results. The blue line refers to our benchmark simulation in better agreement with IceCube data as a function of the local rate of choked GRBs. The yellow line refers to a simulation for a less luminous GRB. The dashed grey line shows the Waxmann & Bahcall upper bound [7].

References

[1] P. Meszaros and E. Waxman, PRL Phys. Rev. Lett. **87**, 171102 (2001)
[2] M. G. Aartsen et al. (IceCube Collaboration), Science **342**, 1242856 (2013)
[3] M. Ackermann et al. (Fermi LAT Collaboration), ApJ **799**, 86 (2015)
[4] M. Fasano, S. Celli, D. Guetta, A. Capone, A Zegarelli and I. Di Palma, JCAP **09**, 044 (2021)
[5] R. Abbasi et al. (IceCube Collaboration), Phys. Rev. D **104**, 022002 (2021)
[6] J. Stettner, PoS **ICRC2019**, 1004 (2019)
[7] J. Bahcall and E. Waxman, Phys. Rev. D **64**, 023002 (2001)

AUTHOR INDEX

AUTHOR INDEX